DISCARD

Spectacular Chemical Experiments

Herbert W. Roesky

1807–2007 Knowledge for Generations

Each generation has its unique needs and aspirations. When Charles Wiley first opened his small printing shop in lower Manhattan in 1807, it was a generation of boundless potential searching for an identity. And we were there, helping to define a new American literary tradition. Over half a century later, in the midst of the Second Industrial Revolution, it was a generation focused on building the future. Once again, we were there, supplying the critical scientific, technical, and engineering knowledge that helped frame the world. Throughout the 20th Century, and into the new millennium, nations began to reach out beyond their own borders and a new international community was born. Wiley was there, expanding its operations around the world to enable a global exchange of ideas, opinions, and know-how.

For 200 years, Wiley has been an integral part of each generation's journey, enabling the flow of information and understanding necessary to meet their needs and fulfill their aspirations. Today, bold new technologies are changing the way we live and learn. Wiley will be there, providing you the must-have knowledge you need to imagine new worlds, new possibilities, and new opportunities.

Generations come and go, but you can always count on Wiley to provide you the knowledge you need, when and where you need it!

William J. Pesce
President and Chief Executive Officer

Peter Booth Wiley
Chairman of the Board

Spectacular Chemical Experiments

Herbert W. Roesky

Foreword by George A. Olah

WILEY-VCH Verlag GmbH & Co. KGaA

The Author

Prof. Dr. Dr. h.c. mult. Herbert W. Roesky
Institute of Inorganic Chemistry
Georg-August-University of Göttingen
Tammannstr. 4
37077 Göttingen
Germany

All books published by Wiley-VCH are carefully produced. Nevertheless, authors, editors, and publisher do not warrant the information contained in these books, including this book, to be free of errors. Readers are advised to keep in mind that statements, data, illustrations, procedural details or other items may inadvertently be inaccurate.

Library of Congress Card No.:
applied for

British Library Cataloguing-in-Publication Data
A catalogue record for this book is available from the British Library.

Bibliographic information published by the Deutsche Nationalbibliothek
The Deutsche Nationalbibliothek lists this publication in the Deutsche Nationalbibliografie; detailed bibliographic data are available in the Internet at <http://dnb.d-nb.de>.

© 2007 WILEY-VCH Verlag GmbH & Co. KGaA, Weinheim

All rights reserved (including those of translation into other languages). No part of this book may be reproduced in any form – by photoprinting, microfilm, or any other means – nor transmitted or translated into a machine language without written permission from the publishers. Registered names, trademarks, etc. used in this book, even when not specifically marked as such, are not to be considered unprotected by law.

Typesetting   SNP Best-set Typesetter Ltd.
Printing   betz-druck GmbH, Darmstadt
Binding   Litges & Dopf GmbH, Heppenheim
Cover Design   WMX-Design, Bruno Winkler, Heidelberg
Wiley Bicentennial Logo
Richard J. Pacifico

Printed in the Federal Republic of Germany
Printed on acid-free paper

ISBN: 978-3-527-31865-0

Foreword

Whoever is ignorant
of the four elements,
of the strength they wield
and of their quality,
cannot master
the band of the spirits.

Johann Wolfgang von Goethe, *Faust I, Study*

In *Faust*, Johann Wolfgang von Goethe shows, in masterly fashion, the magic attraction of the elements or alchemy (chemistry), whilst at the same time claiming that another field – the one of the spirits or, in a more open interpretation, the one of philosophy and arts – is of fundamental importance. In the present collection of spectacular chemical experiments, Herbert W. Roesky has created a fascinating amalgam of brilliant chemical experiments, in addition to a variety of amusing and pensive aphorisms, quotations, anecdotes, and small stories originating from this universe that is almost lost to the scientist or, more generally speaking, to *homo technicus* or at least far away from him. In his book *Chemical Curiosities*, the author has already proved convincingly, that this synthesis of natural science and arts is not a combination of fire and water but rather two sides of the same medal. It is very good that again a bridge has been thrown across two disciplines of the modern world which seem to be far away from each other.

This book contains new "bang and smoke" experiments that make people's hearts beat faster (see Münchhausen's canon ball, bromide and potassium!!). It can also revive playful instincts ("sodium billiards") or raise magic reactions ("the alchemist's gold"). It is possible that some people might prefer the aesthetics of some experiments or the fascination of art (beautiful color experiments). In this book, the varied journey through an easily understandable pure scientific

Spectacular Chemical Experiments. Herbert W. Roesky
Copyright © 2007 WILEY-VCH Verlag GmbH & Co. KGaA, Weinheim
ISBN: 978-3-527-31865-0

universe with anecdotes, quotations, and brief stories introducing every experiment not only becomes an adventure but also perhaps gives us back some of the magic that is inherent in the worlds of both chemistry and arts.

George A. Olah

Contents

Foreword V

Preface XIII

Part I: Water 1

Experiment 1 Spontaneous ignition by adding water 3

Experiment 2 Blowing-up an iron ball 5

Experiment 3 Hydration 7

Experiment 4 Osmosis 11

Experiment 5 Re-gelation of ice 15

Experiment 6 Sugar coal by splitting off water from sugar with sulfuric acid 17

Experiment 7 Sodium billiards 19

Experiment 8 Boiling water in a paper bowl 23

Experiment 9 The density differences of H_2O and D_2O 25

Experiment 10 Fire under water 27

Experiment 11 Safe production of detonating gas 29

Experiment 12 Fuel cell for hydrogen and oxygen 33

Experiment 13 Hydrogen in status nascendi 35

Spectacular Chemical Experiments. Herbert W. Roesky
Copyright © 2007 WILEY-VCH Verlag GmbH & Co. KGaA, Weinheim
ISBN: 978-3-527-31865-0

Experiment 14 Effusion of hydrogen *37*

Experiment 15 Freezing mixture *39*

Experiment 16 Rapid crystallization *41*

Experiment 17 Magic eggs *43*

Experiment 18 Colored kinetics *47*

Experiment 19 Flushing peppermint tea *49*

Experiment 20 Chemiluminescence *51*

Experiment 21 The colors white-yellow-black *53*

Experiment 22 Nitrogen and hydrogen by electrolysis *55*

Experiment 23 Demonstration of the plasma state: "A sparkling cross" *57*

Part II: The color blue *59*

Experiment 24 Witching hour *61*

Experiment 25 Molybdenum blue *63*

Experiment 26 Combustion of sulfur in oxygen *65*

Experiment 27 Phosphorus salt pearl or cobalt salt pearl *67*

Experiment 28 Fehling's solution *69*

Experiment 29 Activated carbon decolorizes water blue *71*

Experiment 30 Blue bottle – The blue miracle *73*

Experiment 31 Generation of blue (N_2O_3) dinitrogen trioxide *75*

Experiment 32 Bleaching with a household product *77*

Experiment 33 Ink blue – solvated electrons *79*

Part III: The color red 81

Experiment 34   Purple or colorless – an entertaining demonstration 83

Experiment 35   A "red component" in newspapers 85

Experiment 36   Bleaching of tomato juice with chlorine on a micro scale 87

Experiment 37   Production of non-drinkable red wine 89

Experiment 38   Red wine as a color indicator 91

Part IV: Colloids, sols, and gels 93

Experiment 39   Silica gel from alkali silicates 95

Experiment 40   Red gold 97

Experiment 41   Red gold sol 99

Experiment 42   Blue gold sol 101

Experiment 43   Cherry red gold sol 103

Experiment 44   The blue gold 105

Experiment 45   Silver sol by electric discharge 107

Experiment 46   How to make a silver sol 109

Experiment 47   The reaction of silver nitrate with tannin 111

Part V: Fascinating experiments by self-organization 113

Experiment 48   Dissipative structures: Chemical patterns in aqueous solution 115

Experiment 49   Acidic acid butyl ester in the presence of bromocresol green 119

Experiment 50   Precipitation using the gas phase 121

Experiment 51 Methods become accepted: Nessler's reagent and gaseous ammonia   *123*

Experiment 52 Reduction of $KMnO_4$ with ethyl alcohol   *125*

Experiment 53 Alcohol test   *129*

Experiment 54 An old hat with new feathers: the precipitation of AgCl with HCl gas   *133*

Part VI: Chemical varieties   *135*

Experiment 55 A chemical buoy   *137*

Experiment 56 Flower power   *139*

Experiment 57 Münchhausen: the flying styrofoam ball   *141*

Experiment 58 The remarkable rocket   *145*

Experiment 59 Eatable burning banana   *149*

Experiment 60 Burning pecan   *151*

Experiment 61 Sparks and shining fire   *153*

Experiment 62 Like magic . . . the reduction of copper oxide   *157*

Experiment 63 Electric current from a beer can   *159*

Experiment 64 Magnesium powder burning in the air   *161*

Experiment 65 The alchemist's gold   *163*

Experiment 66 Imitate a spider   *165*

Experiment 67 Is it methyl alcohol or ethyl alcohol?   *167*

Experiment 68 Oxygen content of the air   *169*

Experiment 69 Rapid rust   *171*

Experiment 70 Shining dry ice   *173*

Experiment 71   Smoke rings   *177*

Experiment 72   Saturn's rings   *181*

Experiment 73   Oxygen from Ag_2O *183*

Experiment 74   Flour dust explosion   *185*

Experiment 75   Bromine and potassium   *189*

Experiment 76   Current-free shining flat-bottomed cylinder   *191*

Experiment 77   Rotating advertising column   *193*

Experiment 78 S_4N_4 – A pick-me-up   *195*

Experiment 79   Thunderclap   *197*

Experiment 80   A heavyweight does not stick to the bottom   *199*

Experiment 81   Icarus and the sun   *201*

Experiment 82   Disposal of sodium and potassium residues   *203*

Part VII: The art gallery of chemistry   *205*

Experiment 83   Color composition: Chemistry is art   *207*

Experiment 84   Underwater dance   *209*

Experiment 85   Blue mist   *211*

Experiment 86   Colorful clouds   *213*

Conclusion   *217*

Index   *219*

Preface

After having delivered more than 150 "experimental" lectures outside Göttingen, and in eight different countries, I decided to write a third book containing spectacular experiments. These experiments are introduced by poems, anecdotes, epigrams, and interesting stories, which elucidate the ubiquitous character of chemistry and arts.

I realized that the existence of the two cultures, as stated by C.P. Snow in 1959 for science and arts, is perceived only after puberty. This is due to the different kinds of education and different schools of thought which are presented to school children. As early as 1947, the mathematician Wiener referred to this difference in his book, *The Intellectual and the Scientist*:

"We have seen, that communication is the mortar of society and that those, who charge themselves to maintain undisturbed the means of communication are mostly responsible for the continuation or the decay of our culture. Unfortunately, these priests of communication are separating in two orders or sects, which defend different principles and have a different education. These two orders of communication priests are, on the one hand, the intellectuals and the arts scholars, on the other hand, the scientists. [. . .] I do not criticise the hostility of the intellectuals and arts scholars towards science and machine age. Hostility is positive and creative, and much of the progression of the machine age demands active and deliberate resistance. I rather criticise him because of his lack of interest in the machine age. He thinks that it is not important to know thoroughly the principles of science and technology, and to become active where these principles are concerned. He is hostile but his hostility does not urge him to do something. It is some kind of homesickness of the past, a vague uneasiness with regard to the present more than any deliberate attitude."

Moreover, I think it is important that chemistry is presented in a charming and inspiring way, and not only with "bang and smoke" experiments. "Chemistry has to be good", people say, and this means positive rather than deterring experiences in the lecture hall. Communication and inspiration between science and arts is also very important.

Spectacular Chemical Experiments. Herbert W. Roesky
Copyright © 2007 WILEY-VCH Verlag GmbH & Co. KGaA, Weinheim
ISBN: 978-3-527-31865-0

The experiments in this book are not presented in any systematic order. Apart from the classical experimental art, two new types of presentation are introduced. On the one hand, these are reactions in the gas phase, for example the precipitation of silver chloride and the identification of alcohol. On the other hand, we use a digital camera in order to record the reaction in pictures; this process can be found in Part VII, the "Art Gallery of Chemistry". Using this method, the person performing these experiments can demonstrate his or her work convincingly, even outside the laboratory, such that science and art – two demanding and creative activities – can be shown together.

During the writing of this book I have greatly appreciated the books of B.Z. Sakhashiri, *Chemical Demonstration – A Handbook for Teachers of Chemistry*, the *Journal of Chemical Education*, and *Chemie in unserer Zeit*. I am also very grateful to Henry Fraatz, who not only supervised and optimized all the experiments but without whom this book would not have been written.

Finally, I would like to quote the advice of Franz Kafka for all those who have decided to perform experiments to fill the audience with enthusiasm:

"Do not spend your time by searching obstacles, maybe there are none."

Herbert W. Roesky

Part I
Water

I don't know who discovered water, but it probably wasn't a fish.

Herbert Marshall McLuhan

From heaven it comes,
To heaven it goes,
And down again
To the earth
Endlessly changing.

Ferdinand Fischer

Water is one of Aristotle's four elements. Thales described it as the only true element from which all the other materials originate. Today, we know that without water, life and evolution are impossible.

This is true even at the beginning of the 21st century. Europe and the United States are competing to discover if there is human habitation on Mars (Beagle 2 and Spirit). In the solar system, Mars is much more like Earth, and scientists consider that, some 3.5 billion years ago, today's icy planet was warm and humid. Although there is always water on Mars, it exists as a more or less thick layer of ice.

Water is vital both inside and outside the human body.

The water percentage in the human body:

Age	Water content (%)
Day of birth	79
5 years	2
16 years	58
Adult	
Normal weight	62
Very slim	69
Very fat	42

Spectacular Chemical Experiments. Herbert W. Roesky
Copyright © 2007 WILEY-VCH Verlag GmbH & Co. KGaA, Weinheim
ISBN: 978-3-527-31865-0

The water percentage inside the different organs:

Organ	Water content (%)
Eyeball	99
Brain	84
Heart	74
Liver	72
Bones	55
Hair	4
Teeth	0.2

Inside the human body, water transports dissolved materials to the different organs and the cells, and it also serves to maintain the functions of the cells. Water is also necessary for the digestion of food and the transportation of its components.

Water is not always drinking water; the oceans, for example, contain 4% of salt which is mostly sodium chloride. Shipwrecked persons who drink water with such a high salt content can survive for only a short period of time.

Water is a very good solvent. In fact, drinking water contains almost every soluble material with which it comes into contact. However, only 0.27% of all of the water on Earth can be used as drinking water.

Experiment 1
Spontaneous Ignition by Adding Water

Whoever is ignorant
of the elements,
of the strength they wield
and of their quality
Cannot master
The band of the spirits.

Johann Wolfgang von Goethe

Apparatus	A fire-proof support, one 250-mL beaker, one wash-bottle, safety glasses, protective gloves.
Chemicals	Wood shavings, Na_2O_2, water (or champagne, beer, etc.).
Attention!	Na_2O_2 reacts almost like sodium spontaneously with water. Na_2O_2 and hydrogen peroxide can cause burns, and skin contact must be avoided. Do not scale up the amount of Na_2O_2. Safety glasses and protective gloves must be used at all times.
Experimental Procedure	The wood shavings are loosely filled into the beaker and the latter is placed on the fire-resistant support. Before starting the experiment, 0.4 g of Na_2O_2 is placed on the wood shavings and immediately a few drops of water are added. The water reacts spontaneously with the Na_2O_2, and the wood shavings start to burn. In most of cases, the beaker cracks.
Explanation	Na_2O_2 is a strong oxidizer and reacts very often explosively with unsaturated organic compounds under incandescence. In the presence of small amounts of water, Na_2O_2 reacts under the elimination of oxygen:

$$Na_2O_2 + H_2O \rightarrow 2\ NaOH + O$$

Spectacular Chemical Experiments. Herbert W. Roesky
Copyright © 2007 WILEY-VCH Verlag GmbH & Co. KGaA, Weinheim
ISBN: 978-3-527-31865-0

The NaOH reacts catalytically under decomposition of the intermediately formed H_2O_2. However, at low temperatures Na_2O_2 reacts with water under formation of NaOH and H_2O_2:

$$Na_2O_2 + 2\ H_2O \rightarrow H_2O_2 + 2\ NaOH$$

Experiment 1: A beaker with burning wood shavings and sodium peroxide.

Experiment 2
Blowing-Up an Iron Ball

We ourselves are the measure of the miraculous, when we would look for a general measure, then the miraculous would disappear, and everything would have the same size.

Georg Christoph Lichtenberg

Apparatus	An iron ball equipped with a screw lock (Phywe Göttingen), one PVC trough (30 cm in diameter and 15 cm inside height), safety glasses, protective gloves.
Chemicals	Wet ice, salt for the cooling bath, ice water.
Attention!	Safety glasses and protective gloves must be used at all times.
Experimental Procedure	The iron ball is filled with ice water and the screw lock is tightly closed. The iron ball is then placed in the already prepared freezing mixture (ice and sodium chloride). After about 20–30 minutes, the iron ball ruptures under a moderate explosion. Sometimes, ice and parts of the ball are thrown out of the trough.
Explanation	During the transition from the liquid to the solid state, water expands its volume by 9%. This transition from liquid to solid state while reducing the density is a very rare case for liquids. On the one hand, this phenomenon is an important property in Nature, since when water freezes the ice covers the surface of the water and protects the living organisms beneath. On the other hand, water increases surface weathering by breaking loose stones, whereupon the cracks are filled with water and freeze in winter.

6 | Experiment 2 Blowing-Up an Iron Ball

Experiment 2: The iron ball, before and after being blown up.

Experiment 3
Hydration

Science when understood correctly, cures man of his pride, thus showing him his limits.

Albert Schweizer

Apparatus — A temperature-indicating instrument or a thermometer, two thermoelements, two 250-mL beakers, and two glass rods. A spatula should be used for the chemicals, two stands, clamps, bosses.

Chemicals — Dry $CaCl_2$, $CaCl_2 \cdot 6H_2O$, distilled water.

Attention! — Safety glasses and protective gloves must be used at all times.

Experimental Procedure — The thermoelement is attached at the stand and almost reaches the bottom of the beaker. The temperature-indicating instrument is switched on to show room temperature. To the first beaker is given seven spatula loads of dry $CaCl_2$, and then about 50 mL of water is added, with stirring. The temperature-indicating instrument shows an increase in temperature to 35–40 °C.

The second reaction is repeated with $CaCl_2 \cdot 6H_2O$ under the same conditions; this results in a final temperature level of 8–10 °C.

Explanation — The solution enthalpy (ΔH_L) depends on the lattice energy (ΔH_u) and the hydration enthalpy (ΔH_H) according to the following equation:

$$\Delta H_L = \Delta H_u - \Delta H_H$$

where: $L \cong$ solution enthalpy; $U \cong$ lattice enthalpy; and $H \cong$ hydration enthalpy.

When $H > U$, then L becomes negative, and during the dissolution of the salt the solution will be warmed up. In contrast, when $H < U$ the solution will be cooled down.

Spectacular Chemical Experiments. Herbert W. Roesky
Copyright © 2007 WILEY-VCH Verlag GmbH & Co. KGaA, Weinheim
ISBN: 978-3-527-31865-0

The lattice of $CaCl_2$ consists of calcium cations and chloride anions. When $CaCl_2$ is dissolved in water, an increase in temperature is observed.

$$CaCl_2 + 18\ H_2O \rightarrow Ca(H_2O)_6^{2+} + 2\ Cl(H_2O)_6^{-}$$

Here, the change of ΔH must be larger than that of ΔU. But when $CaCl_2 \cdot 6H_2O$ is dissolved in water the system cools down.

$$CaCl_2 \cdot 6\ H_2O + 12\ H_2O \rightarrow Ca(H_2O)_6^{2+} + 2\ Cl(H_2O)_6^{-}$$

In the case of $CaCl_2$ the hydration enthalpy is larger, while the Ca^{2+} and Cl^{-} have to be hydrated. In the lattice of the water-containing $CaCl_2 \cdot 6H_2O$, the Ca^{2+} ions are already present as an aqua complex of composition $Ca(H_2O)_6^{2+}$. Therefore, the hydration enthalpy is lower compared to that of the dry $CaCl_2$. Consequently, H is < U and the dissolution process results in a cooling down of the solution.

Waste Disposal

The waste solutions can be poured down the drain.

Experiment 3 Hydration | 9

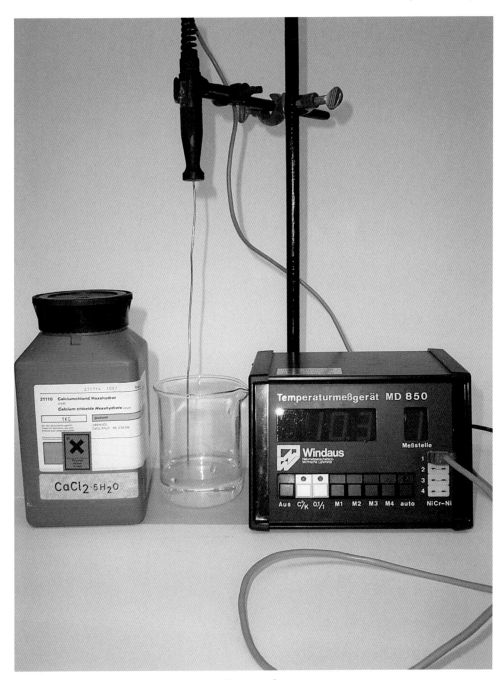

Experiment 3: The set-up for measuring temperature changes.

Experiment 4
Osmosis

If nature had as many laws as the state, even God could not be organized.

Ludwig Börne

Apparatus

One Pfeffer's cell consisting of a bell jar with a volume of 100 mL and a screw connection equipped with a gasket. The lower part of the bell jar contains a plane-ground joint covered with a cellophane membrane and a supporting facility. One capillary tube of outer diameter 6 mm, inner diameter 1.2 mm, and length 1.5 m. A 1-L beaker, a laboratory jack, stand, clamps, bosses, protective gloves, safety glasses.

Chemicals

A concentrated sodium chloride solution or a concentrated sugar solution is prepared, using distilled water. For the demonstration the solution is colorized with fluorescein or methyl red.

Attention!

Safety glasses and protective gloves must be used at all times.

Experimental Procedure

Before the experiment is carried out, the cellophane membrane is attached to the bell jar by means of the supporting facility. The bell jar is completely filled with the colorized solution and the capillary tube is attached using the screw connection. The screw connection of the bell jar is used as a point of attachment with a stand. The beaker is filled two-thirds full with water and then slowly lifted, using the laboratory jack, until half of the bell jar is in the water. Initially, the level of the colorized solution in the capillary is 1–2 cm above the screw connection. After a short time, the solution begins to rise in the capillary, and after 60 minutes it reaches 1 meter in height.

Explanation

At a semi-permeable membrane the osmotic pressure or the remaining hydrostatic pressure can be measured. Here, the membrane is permeable for water but not for the sugar molecules. According to van't Hoff, the osmotic pressure of a diluted solution equals that of

Spectacular Chemical Experiments. Herbert W. Roesky
Copyright © 2007 WILEY-VCH Verlag GmbH & Co. KGaA, Weinheim
ISBN: 978-3-527-31865-0

a gas phase, when the dissolved molecules occupy the same volume and have the same temperature, like an ideal gas. Consequently, the molar mass of dissolved molecules can be determined by measuring the osmotic pressure.

Osmosis is important for all living beings. In cells, the cytoplasm is surrounded by semi-permeable membranes, which are important for balancing the amount of water and the dissolved substances. In human blood, the osmotic pressure ranges from 70 000 to 80 000 Pa; this is equal to the osmotic pressure of a 0.95% solution of sodium chloride (isotonic solution). A more highly concentrated solution of sodium chloride would lead to an involution of the blood cells, whereas a lower concentration corresponds to a lower osmotic pressure and would result in a swelling of the blood cells. The latter procedure is known as plasmolysis.

Waste Disposal

The solutions can be poured down the drain.

Experiment 4: Pfeffer's cell, with capillary.

Experiment 5
Re-Gelation of Ice

The elements are almost uncontrolled: the earth tries to seize the water and force it to solidify forming earth, rock or ice, to stay in earth's boundary.

Johann Wolfgang von Goethe

Apparatus A wooden rack consisting of a bottom plate (20 cm long and 10 cm wide) with two vertical boards (30 cm high and 10 cm wide). The two vertical boards have at their upper ends a notch to carry the ice block. A metal wire of 0.2 mm diameter, two weights each of 1 kg, a trough to collect the water from the melting ice, protective gloves, safety glasses.

Chemicals An ice block which is 25 cm long, 10 cm wide and 2 cm deep is prepared in a flat plastic container in a freezer.

Attention! Safety glasses and protective gloves must be used at all times.

Experimental Procedure The ice block is placed on edge in the wooden rack, within the notches. The two weights are connected with the metal wire approximately 30 cm long. The metal wire with the weights is left to hang on the ice. Due to the pressure of the wire on the ice, the wire begins to move through the ice, from top to bottom. This takes about 30–40 minutes. However, as the wire moves through the ice, the ice above the wire is re-formed, such that the ice block is not cut into two pieces.

Explanation The melting point of a substance depends on the pressure applied. In the case of water, an increase in pressure lowers the melting point. Therefore, below the wire the ice melts, but above the wire the ice is formed again.

Spectacular Chemical Experiments. Herbert W. Roesky
Copyright © 2007 WILEY-VCH Verlag GmbH & Co. KGaA, Weinheim
ISBN: 978-3-527-31865-0

Experiment 5: The re-gelation of ice.

Experiment 6
Sugar Coal by Splitting off Water from Sugar with Sulfuric Acid

Only those who do not go in search of things are sure not to make mistakes.

Albert Einstein

Apparatus	One 250-mL beaker, a glass rod, protective gloves, safety glasses.
Chemicals	Sugar, concentrated sulfuric acid.
Attention!	Concentrated sulfuric acid is highly corrosive, and causes severe burns to skin and eyes. Rubber gloves must be used. The experiment must be carried out in a well-ventilated fume hood. Safety glasses and protective gloves must be used at all times.
Experimental Procedure	A 50-g quantity of sugar is placed in the beaker, and as much sulfuric acid added such that the mixture can be well stirred with the glass rod (the sulfuric acid level should be 1 cm above the level of the undissolved sugar). After 2–3 minutes of stirring, the mixture turns black, and warms up under foam formation and the elimination of water vapor. A porous, solid sugar "coal" has been formed.
Explanation	Treatment of sugar with concentrated sulfuric acid results in water elimination and decomposition of the sugar under the formation of porous carbon. The equation shows the decomposition of sugar under formation of carbon and water: $$C_mH_{2n}O_n \rightarrow mC + nH_2O$$
Waste Disposal	The sugar coal is washed with water to remove the adhering sulfuric acid. The waste water can then be flushed down the drain.

Spectacular Chemical Experiments. Herbert W. Roesky
Copyright © 2007 WILEY-VCH Verlag GmbH & Co. KGaA, Weinheim
ISBN: 978-3-527-31865-0

Experiment 6 Sugar Coal by Splitting off Water from Sugar with Sulfuric Acid

Experiment 6: The production of sugar coal.

Experiment 7
Sodium Billiards

I think that instruction and the study of the history of sciences are indispensable . . . our textbooks are deficient in this regard.

Richard Willstätter

When everybody is thinking the same, nobody is thinking much.

Ralph Lippmann

Apparatus One round glass dish (14 cm diameter, inner height 3 cm), one rectangular glass rack with four glass stands made from round glass rods, each of which is 9 cm long, and a total height (including the stands) of ca. 2 cm (see illustration), knife or spatula, a pair of tweezers, filter paper, protective gloves, safety glasses.

Chemicals Sodium, washing-up liquid, water, 1% phenolphthalein solution, potassium metal.

Attention! Sodium and potassium react with water EXPLOSIVELY! No more than 200 mg of sodium should be used.

Sodium is often used to dry solvents such as ether, hydrocarbons, and tertiary amines. It is also used for reduction reactions of metal and non-metal halogenides. Sodium is dangerous when it comes into contact with water. The reaction with water is exothermic, and produces hydrogen; when in contact with air, explosive oxihydrogen is generated.

Safety glasses and protective gloves must be used at all times.

Experimental Procedure A small piece of sodium is cut from a bar, using a knife or a spatula. The mineral oil is removed from the sodium using the filter paper. Before starting the experiment, the sodium bar must be placed back in the bottle with mineral oil and the bottle must be closed with a stopper before carrying out the experiment.

Spectacular Chemical Experiments. Herbert W. Roesky
Copyright © 2007 WILEY-VCH Verlag GmbH & Co. KGaA, Weinheim
ISBN: 978-3-527-31865-0

The rectangular glass rack is placed into the glass dish. The water is then added slowly so that the glass rods are covered with water up to half of their diameter. One drop of washing-up liquid is then added (to reduce the surface tension of the water), and the small piece of sodium is placed with the tweezers into the rectangular glass rack. The water reacts with the sodium and enough heat is produced to melt the sodium.

This experiment can also be performed with 50 mg of potassium; however, the reaction is much more violent, and generally an explosion of the rapidly formed hydrogen occurs.

Explanation

The liquid sodium moves around inside the rectangle like a "billiard ball". The sodium does not stick to the side of the dish because of the round glass rods and the reduced surface tension of the water. It also means that there is not enough heat to produce an explosion (this often happens when a beaker is used and the sodium sticks to the side of the beaker). When some drops of phenolphthalein solution are added, the color changes from colorless to red, showing the formation of sodium hydroxide:

$$2\text{ Na} + 2\text{ H}_2\text{O} \rightarrow 2\text{ NaOH} + \text{H}_2$$

Waste Disposal

The alkaline solution can be flushed down the drain.

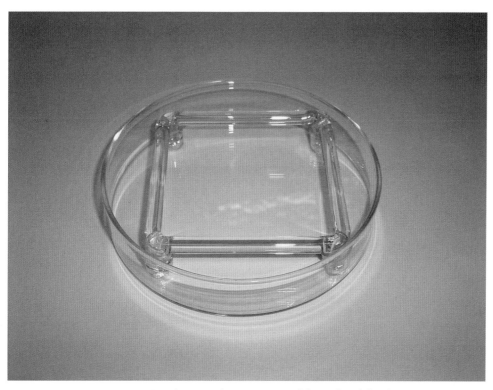

Experiment 7: The apparatus used for "sodium billiards".

Experiment 8
Boiling Water in a Paper Bowl

Water was the origin of the world and of all its creatures.

Paracelsus

When I want to play I make a net out of the degrees of longitude and latitude, and use it to catch whales in the Atlantic Ocean.

Mark Twain

Apparatus	A ring stand, three paper clips, writing paper, Bunsen burner, scissors, protective gloves, safety glasses.
Chemicals	Water.
Attention!	Safety glasses and protective gloves must be used at all times.
Experimental Procedure	The paper is placed on the ring stand, making a bowl, and attaching it with the paper clips. Any overlapping paper is cut off (otherwise this could catch fire). The bowl is filled with approximately 100 mL of water which is then heated with a Bunsen burner. The water becomes hot, but the paper does not burn.
Explanation	The heat of the flame is conducted through the paper to the water by slowly increasing the temperature of the water to its boiling point.
Reference	L.A. Ford, *Chemical Magic*, Dover Publications, Inc., New York, 1993.

Experiment 9
The Density Differences of H₂O and D₂O

Hydrogen bonds are characteristic for the structure of water in the liquid and the solid phase. Already in 1939, Linus Pauling predicted the importance of the hydrogen bond:
"Although the hydrogen bond is not a strong bond, it has great significance in determining the properties of substances. Because of its small bond energy and the small activation energy involved in its formation and rupture, the hydrogen bond is especially suited to play a part in reactions occurring at normal temperatures. It has been recognized that hydrogen bonds restrain protein molecules to their native configurations, and I believe that as the methods of structural chemistry are further applied to physiological problems, it will be found that the significance of the hydrogen bond for physiology is greater than that of any other single structural feature."

Linus Pauling, *The Nature of the Chemical Bond*

Apparatus	One arc lamp or overhead projector and a mirror, a cuvette ($5 \times 5 \times 2.5 \text{ cm}^3$), protective gloves, safety glasses.
Chemicals	Water, heavy water (deuterium oxide).
Attention!	Safety glasses and protective gloves must be used at all times.
Experimental Procedure	Before carrying out the experiment in the lecture hall, ice cubes of H_2O and D_2O respectively are prepared in a plastic container in a freezer. The cuvette filled with water is projected onto a screen by an arc lamp. The addition of an ice cube of normal water shows, like expected, that this ice cube is floating on the water, whereas the cube of D_2O sinks to the bottom of the cuvette.
Explanation	Normal ice has a density of 0.9 g mL^{-1} at $-20\,°C$, whereas ice of D_2O has a density of 1.1 g mL^{-1} at the same temperature.
Reference	A.B. Ellis, E.A. Adler, F.H. Juergens, *J. Chem. Educ.* **1990**, *67*, 159.

Spectacular Chemical Experiments. Herbert W. Roesky
Copyright © 2007 WILEY-VCH Verlag GmbH & Co. KGaA, Weinheim
ISBN: 978-3-527-31865-0

Experiment 10
Fire Under Water

The utmost pleasure of a scientist is the usefulness of a discovered truth by an inquiring mind.

Jacobus Henricus van't Hoff

Apparatus	A 500-mL goblet, 30-mL pipette, protective gloves, safety glasses.
Chemicals	White phosphorus, $KClO_3$, concentrated H_2SO_4, distilled water.
Attention!	White phosphorus is highly toxic and pyrophoric, even at room temperature. $KClO_3$ is a strong oxidant. Contact with organic materials must be avoided. Safety glasses and protective gloves must be used at all times.
Experimental Procedure	3 g of $KClO_3$ are placed in a goblet and covered with a 3-cm layer of distilled water. A piece of white phosphorus (0.2–0.3 g) is added. **By no means should the white phosphorus be added to the $KClO_3$ before adding the water. Danger of explosion!** When the sulfuric acid is added drop-by-drop close to the piece of phosphorus, lightning flashes are observed. If the reaction is too weak, the sulfuric acid must be added more quickly. However, if the sulfuric acid is added too quickly, the piece of white phosphorus melts and burns at the surface of the water. This experiment is especially exciting when the reaction is performed in an almost dark room!!
Explanation	The sulfuric acid reacts with $KClO_3$ to give the $HClO_3$, which decomposes according to the following equation: $3\ HClO_3 \rightarrow HClO_4 + 2\ ClO_2 + H_2O$ The ClO_2 oxidizes the phosphorus under incandescence.

Spectacular Chemical Experiments. Herbert W. Roesky
Copyright © 2007 WILEY-VCH Verlag GmbH & Co. KGaA, Weinheim
ISBN: 978-3-527-31865-0

Waste Disposal Phosphorus residues are treated with aqueous copper sulfate to produce the corresponding phosphide, which is oxidized by a strong alkaline sodium hypochlorite solution. Finally, milk of lime is added to give calcium phosphate and copper hydroxide. The resulting solution and deposits are collected in the container for heavy metals.

Experiment 11
Safe Production of Detonating Gas

The gas which results from the reaction of metals such as zinc or iron with acids was probably already known by Paracelsus (1493–1541), who had observed that a gas is produced when he dissolved iron in vitriol. "Air is produced and bursts out like a storm." However, he did not realize that this gas was inflammable. This was observed first by Turquet de Mayenne (1573–1655) when he reacted sulfuric acid with iron. Finally, Sir Henry Cavendish was able to show that the reaction between hydrogen and oxygen produces detonating gas.

Heumann describes how he produced detonating gas electrolytically:

"A wide-necked small jar is closed by a cork stopper carrying a bent gas delivery tube which is open at both ends; the cork stopper carries also two glass tubes in which there are two platinum wires. The small tubes which have been reduced before in diameter at the end are molten airtight to the platinum wires, which can be done by a simple candle flame.
The jar is filled almost completely with sulfuric acid which has been diluted with twelve times the weight of water, and then the platinum wires are connected to the poles of a galvanic battery made out of several cells.
The first generated gas contains still the air inside the apparatus; when we may assume that the detonating gas is pure, the gas exit tube is put into a soap solution containing some glycerine placed in a bowl, and then the soap bubbles are lighted after having removed them from the gas generator. This apparatus has to be protected carefully of being close to a flame (or a platinum sponge), otherwise dangerous explosions may occur. Because the electrical current becomes very weak when it is passing through the acidified water, the platinum wires have to be bent so that the sheet strips are not too far away from each other. However, it is important that during the experiment the platinum wires or the sheet strips inside the jar are not in direct contact, because then, even when the galvanic current is low, they will become very hot or even glow, and this may cause an explosion of the detonating gas inside the apparatus. Instead of passing the gas into soap water it is possible to collect it in a small tube and ignite it with a platinum sponge."

Spectacular Chemical Experiments. Herbert W. Roesky
Copyright © 2007 WILEY-VCH Verlag GmbH & Co. KGaA, Weinheim
ISBN: 978-3-527-31865-0

Many severe accidents occurred when preparing detonating gas, because mostly too much detonating gas is prepared. An example:

Flask exploded: Eight school children hurt.

ebb. Buchholz

Eight children of a school in Buchholz (district Harburg) have been hurt on Saturday by an explosion during a chemistry course. The teacher made an experiment with zinc and acid. The flask burst due to the built up pressure and the children were hurt by splinters flying around. One child had to stay in hospital.

Apparatus

4-mL test tube with a screw cap and a septum of silicon/Teflon, two metal paper clips, two cables, two alligator clips, modeling clay, one adjustable transformer with rectifier, 10-mL glass syringe, one lighter, one porcelain dish (ca. 5 cm diameter), one wooden splint, one Pasteur pipette, one stand, two clamps, two bosses, safety glasses, protective gloves.

Chemicals

1 M sulfuric acid, 2 M sodium hydroxide.

Attention!

Mixtures of hydrogen and oxygen are EXPLOSIVE. Safety glasses and protective gloves must be used at all times.

Experimental Procedure

First, all of the modeling clay is placed into the 4-mL tube so that it is 2–4 mm high, and compressed equally. Then, 3 mL of sulfuric acid or sodium hydroxide of the above-specified concentration are added. After closing the tube with the screw cap and the septum, two metal paper clips (to serve as electrodes) are introduced through the septum into the electrolyte; the clips are then connected to the adjustable transformer and the rectifier, using the alligator clips and cable. It is important that the electrodes do not touch each other! At the lower end they should reach deeply into the modeling clay.

The tube and the syringe are fixed onto the stand. The cannula of the syringe is then introduced into the gas room of the tube, and the syringe is fixed vertically above the tube. ONLY THEN MAY THE CURRENT BE SWITCHED ON. The voltage is adjusted so that a moderate gas stream is produced, and the detonating gas is collected in the 10-mL syringe.

The plunger of the syringe can be seen clearly to rise, and when sufficient detonating gas has been collected inside the syringe the current is shut off. The syringe is then removed from the apparatus and the detonating gas carefully injected into the soap solution in the porcelain dish. The small soap bubbles produced may then be ignited with the burning wooden splint. A bang can be clearly heard. When the detonating gas is collected for a second time, the bang of the soap bubbles is much louder compared to the first time (this is because on the first collection some air had been left in the small tube).

Explanation The stoichiometric amount of hydrogen and oxygen for the preparation of detonating gas is best generated by electrolysis of water. The gas mixture is kinetically stable. However, when this mixture is touched by a spark, a flame or a catalytic amount of fine platinum powder, a chain reaction occurs that results in an explosion.

Waste Disposal The sulfuric acid in the small tube is neutralized with sodium hydroxide and flushed down the drain.

Reference Karl Heumann, Instructions for experiments in experimental lectures. Braunschweig, 1876.

Experiment 12
Fuel Cell for Hydrogen and Oxygen

In his "Opening speech about lectures in experimental chemistry" (1852), *Justus von Liebig* formulated clearly how important are experiments and the art of observing in science education:

"There is no art as difficult as the art of observation: a rational intelligent mind together with educated experience, only attainable through practice, are indispensable: not the observer who only visually perceives the object, but rather the observer who sees the different parts and how these parts are connected to the whole. Some are inattentive and overlook half of what they should see, others think they see more than they really do, mixing up what they see with what they think they see, others again see the parts of the whole object but mistakenly connect pieces that should be separated. – The observer who looks at a clock sees not only the pendulum moving to and fro, the clock-face and the moving hands, a child can do that, the observer sees the parts of the clock and moreover, how the hanging weight is connected to the wheelwork and the pendulum to the moving hands. – If the observer has understood what he sees, and if he is capable of combining the conditions, then he attempts to create the observed phenomenon himself through experimentation. The experiment reflects the accuracy of the observation. Performing a series of experiments often means taking an idea apart and testing the correctness of different aspects. The natural scientist conducts experiments in order to prove the accuracy of his interpretation, and to demonstrate all aspects of a phenomenon. If he is able to show that a series of phenomena are the result of the same cause he ends up with a simple result, a so-called natural law. A simple property is called a natural law if it can be used to explain one or more natural phenomena."

Apparatus Two graphite plates (11 cm long, 10 cm wide, 1 cm deep), one electrolytic trough made from glass (16 cm long, 16 cm wide, 5 cm deep), two alligator clips, two cables, a 2-V motor with a rotating disc, adjustable transformer with rectifier, stop watch, stand, bosses, clamps, safety glasses, protective gloves.

Chemicals 0.1 M sulfuric acid.

Spectacular Chemical Experiments. Herbert W. Roesky
Copyright © 2007 WILEY-VCH Verlag GmbH & Co. KGaA, Weinheim
ISBN: 978-3-527-31865-0

Experiment 12 Fuel Cell for Hydrogen and Oxygen

Attention! Safety glasses and protective gloves must be used at all times.

Experimental Procedure Both plates (held by clamps) are placed in the electrolytic trough, and the trough is half-filled (ca. 350 mL) with 0.1 M sulfuric acid. The plates are then connected to the adjustable transformer using the cables. The plates must be made rough by electrolyzing them for 10 seconds with a 20-V direct current. After depolarization, the plates are electrolyzed again for 20 seconds at 20 V. The transformer is then switched off, and the cables are removed and connected to the motor with the rotating disc. The disc is allowed to rotate for least 5–8 minutes.

Even if the electrolysis is longer than 40 seconds, the motor does not move for a longer time. Neither does the concentration of the sulfuric acid (higher or lower) improve the rotation time.

Explanation During the electrolysis, oxygen is formed at the anode, and hydrogen at the cathode. Both gases are absorbed by the carbon electrodes. When the electrolysis is stopped, the electrons of the hydrogen move inside the wire to the anode. This causes a potential difference, which can be measured as a voltage that becomes "visible" by the rotation of the disc. Electron transfer inside the electrolyte is very slow compared to that in the wire.

Waste Disposal The sulfuric acid is neutralized with sodium hydroxide and then flushed down the drain.

Experiment 13
Hydrogen in Status Nascendi

More and more I am convinced that all the progress we may make in the future is depending on the advancement of the scientific education by which one day everybody should learn already at school how to observe and how to think clearly and without any prejudice.

Heinrich Roessler, *Memoirs*

Apparatus	4-mL cuvette with a clamp, a pair of tweezers, spatula, Pasteur pipette, safety glasses, protective gloves.
Chemicals	WO_3, zinc granulate, 12 M hydrochloric acid, distilled water.
Attention!	Hydrochloric acid is highly corrosive and causes severe burns to skin and eyes. Safety glasses and protective gloves must be used at all times.
Experimental Procedure	Within the cuvette, solid WO_3 (ca. 50 mg) is suspended in 1 mL of water. One piece of granulated zinc and 0.5 mL hydrochloric acid are then added. Initially, the yellow-green suspension turns blue and then dark blue. If the experiment is performed with hydrogen from the gas cylinder or the Kipp's apparatus, no color change is observed.
Explanation	When hydrogen is used from Kipp's apparatus, there is no reduction observed due to the high bonding energy of the H_2 molecule. However, when hydrogen in status nascendi is generated, the hydrogen atoms reduce WO_3 to blue $W_4O_{10}(OH)_2$ (charge-transfer-oxide-hydroxide).
Waste Disposal	After neutralization with sodium hydroxide the residues in the cuvette are disposed of in the container for heavy metals.

Spectacular Chemical Experiments. Herbert W. Roesky
Copyright © 2007 WILEY-VCH Verlag GmbH & Co. KGaA, Weinheim
ISBN: 978-3-527-31865-0

Experiment 14
Effusion of Hydrogen

How is it possible to think that science is dry?
Is there anything more beautiful than the unchangeable rules which reign the world, anything more wonderful than the human mind, which is able to discover them?
How unsubstantial, how less imaginative seem to be novels and fairy tales compared to these extraordinary phenomena related to each other by these harmonious rules.

Marie Curie

Hydrogen has high effusion ability. Effusion means the transition of a substance from one medium into another one. Hydrogen can easily pass through porous walls. This effusion ability is related to the density respectively molar mass and the molecular velocity.

Thomas Graham (1805–1869) explained that the relationship between the effusion rate of two gases is inverse to the square root of the molar masses:

$$\frac{r_1}{r_2} = \frac{v_1}{v_2} = \sqrt{\frac{M_2}{M_1}}$$

The relationship of the effusion rate r_1/r_2 is equal to the relationship of the molecular velocity (v_1/v_2).

Apparatus Two stands, two bosses, two clamps, clay cylinder (6 cm diameter, 15 cm long), perforated rubber stopper to fit the glass tube, 600-mL beaker, glass tube, wash bottle with attached capillary, PVC tubing, safety glasses, protective gloves.

Chemicals Hydrogen in a cylinder with reducing valve, saffron, water.

Attention! When experimenting with hydrogen, flames, metal catalysts and sparks should be avoided! Safety glasses and protective gloves must be used at all times.

Spectacular Chemical Experiments. Herbert W. Roesky
Copyright © 2007 WILEY-VCH Verlag GmbH & Co. KGaA, Weinheim
ISBN: 978-3-527-31865-0

Experiment 14 Effusion of Hydrogen

Experimental Procedure

The clay cylinder is fixed vertically to the stand, so that the open side of the cylinder is to the bottom. The clamp should be placed close to the open side of the cylinder. The perforated rubber stopper has to fit the glass tube. The wash bottle and the glass tube are then connected with the PVC tube. The wash bottle is filled with water colored with saffron and mounted in the reverse arrangement. It is then attached to the second stand and a clamp. The beaker is placed over the clay cylinder, so that an adequate air volume can develop between the beaker and the cylinder. When a hydrogen stream is passed into the space between beaker and clay cylinder at the lower part of the beaker, the colored water is pushed strongly out of the wash bottle after a few seconds. The water jet can be seen very clearly because of the saffron coloring.

Explanation

Due to its strong effusion ability, hydrogen penetrates rapidly through the wall of the clay cylinder. This increases the pressure inside the cylinder and results in a water jet.

Waste Disposal

The colored water is flushed down the drain.

Experiment 15
Freezing Mixture

On truth, the essential and foremost condition of erudition, depends everything. Usefulness is only of secondary importance.

Immanuel Kant

Apparatus

A 250-mL Erlenmeyer flask, thermometer (+38 °C to +50 °C), a thin wooden plate (15 × 15 × 2 cm), two 250-mL beakers, safety glasses, protective gloves.

Chemicals

NH_4Cl, $Ba(OH)_2 \cdot 8 H_2O$, (NH_4NO_3).

Attention!

Barium salts are very toxic. Care should be taken to avoid ingestion during all handling operations. Safety glasses and protective gloves must be used at all times.

Experimental Procedure

32 g $Ba(OH)_2 \cdot 8 H_2O$ and 11 g NH_4Cl are weighed separately into the two beakers. Both salts are then transferred into the Erlenmeyer flask, and the mixture is well stirred. The thermometer shows that within 1–2 minutes the temperature falls from ca. 20 °C to –25 °C, and finally to –30 °C. The salt mixture becomes liquid and the flask turns icy white at the outside. The temperature of –20 °C is maintained for several minutes. When the flask is placed on a wet wooden plate, it freezes onto the plate. It is even possible to lift the Erlenmeyer flask together with the plate onto which it is sticking, in order to show the freezing effect.

A simple procedure to generate "cold" is the reaction of NH_4NO_3 with ice water. Here, 14 g of NH_4NO_3 are added to 30 mL of H_2O. The mixture is then stirred for about 1 minute, during which time the temperature falls to –10 °C.

Explanation

The reaction of the two solids proceeds according to the following equation:

$$Ba(OH)_2 \cdot 8 H_2O_{(s)} + 2 NH_4Cl_{(s)} \rightarrow BaCl_2 \cdot 2 H_2O_{(s)} + 2 NH_{3(g)} + 8 H_2O_{(l)}$$

Spectacular Chemical Experiments. Herbert W. Roesky
Copyright © 2007 WILEY-VCH Verlag GmbH & Co. KGaA, Weinheim
ISBN: 978-3-527-31865-0

Shortly after having brought together the two mixtures, a slight smell of ammonia may be noted.

The strong endothermic reaction is entropy controlled. The order of reaction decreases, and the entropy increases strongly. Therefore, the free enthalpy (ΔG) will be negative, because the numerical value of the product $T \times \Delta S$ is higher than the enthalpy, and this is shown by the decreasing temperature.

Waste Disposal

Barium salts are toxic and must be disposed of via the container used for toxic inorganic salts.

Experiment 16
Rapid Crystallization

It is easy for the sugar to be sweet and for the nitrate to be a salt.

Ralph Waldo Emerson

Let us examine a crystal . . . we like the similarity of the sides; the one of the angles doubles our pleasure.

Edgar Alan Poe

The crystal is a chemical cemetery.

Leopold Ruzieka

In his book *Traité élémentaire de Chimie*, Lavoisier describes the rule of the conservation of matter:

Nothing is formed, not in art nor in nature, and a principle states that before and after every process the amount of matter is the same.

The formation of crystals is important, but sometimes difficult. One common use of the crystallization process is that of purification. The growth of single crystals for structural analysis may be very difficult, however.

Apparatus	Glass cuvette ($8 \times 6 \times 2\,cm^3$), 100-mL beaker, Bunsen burner, wire gauze, funnel, folded filter, boiling stone, tripod ring stand, safety glasses, protective gloves.
Chemicals	Lead(II) chloride, distilled water.
Attention!	Lead salts are very toxic. Safety glasses and protective gloves must be used at all times.
Experimental Procedure	2 g of $PbCl_2$ are placed into the 100-mL beaker followed by 50 mL of water, and three boiling stones. The solution is then heated to boiling point, after which the hot solution is filtered directly into the cuvette.

Spectacular Chemical Experiments. Herbert W. Roesky
Copyright © 2007 WILEY-VCH Verlag GmbH & Co. KGaA, Weinheim
ISBN: 978-3-527-31865-0

After a short time (1 minute), crystals appear which grow well as the solution is cooled down.

Explanation

$PbCl_2$ crystallizes in the space group Pmnb. Within its structure, the lead is surrounded by nine chlorine atoms. Six of the chlorine atoms are placed at the corners of a trigonal prism, while the other three atoms are arranged above the three prism planes.

Waste Disposal

The saturated $PbCl_2$ solution is evaporated to dryness and the resultant $PbCl_2$ and crystals from the cuvette are placed back into the stock bottle. The undissolved $PbCl_2$ on the filter paper is dried in air and then also placed into the stock bottle.

Experiment 17
Magic Eggs

Dr. Archie observed her thoughtfully, as if she were a glass full of chemical reactions.

Willa Cather, The Song of the Lark

When I want to resolve science problems, first of all I make some experiments, because I intend to confront myself with the problem after having made the experience, and then to prove why the material is forced to act in the indicated manner.

Leonardo da Vinci (Kenneth Clark, rororo Monographie 50/53)

Apparatus	Two 1-L measuring cylinders, safety glasses, protective gloves.
Chemicals	Two eggs (ca. 65–70 g in weight), NaCl, 6 M HCl, distilled water.
Attention!	Hydrochloric acid can cause burns to skin and eyes. Safety glasses and protective gloves must be used at all times.
Experimental Procedure	The first cylinder contains 500 mL of saturated sodium chloride solution and 500 mL of distilled water. The second cylinder contains 80 mL of HCl and 1000 mL of distilled water.

The first cylinder is filled with 500 mL of saturated sodium chloride solution, and carefully – holding the cylinder in a slanting position – layered with distilled water, until the volume of 1 L is reached. From the outside, the solution inside the cylinder appears homogeneous.

When an egg is placed on top of the solution, it moves down until it reaches the phase boundary where the saturated salt solution begins. In this experiment, the weight of the egg is not important.

The second cylinder is then filled with 80 mL of 6 M HCl and 1000 mL of H_2O. When the egg (70 g) is placed on top of the solution, it moves down to the bottom of the cylinder. After 3–4 minutes, the egg moves up slightly, but then sinks again. This process is repeated several times, with the egg always moving higher upwards in the cylinder. After about 20 minutes, the egg stays at the upper part of the cylinder, at which time the experiment is complete.

Spectacular Chemical Experiments. Herbert W. Roesky
Copyright © 2007 WILEY-VCH Verlag GmbH & Co. KGaA, Weinheim
ISBN: 978-3-527-31865-0

Explanation

The deciding factor of the experimental procedure is the weight of the egg and the concentration of the hydrochloric acid. With an egg weight of 70 g, 70 mL of 6 M hydrochloric acid and 1000 mL of water, the experiment begins to start after 10 minutes and finishes after about 50 minutes. When the egg weighs 65 g, and 50 mL of 6 M hydrochloric acid and 1000 mL of water are used, the egg moves up after 2–3 minutes, and the experiment is completed after 15–20 minutes. With an egg weight of 60 g, 40 mL of 6 M hydrochloric acid and 1000 mL of water, the buoyancy begins after 8 minutes and the experiment keeps going for about 10 minutes.

A problem sometimes occurs when the egg moves up too quickly or too slowly, depending on the amount of acid. Therefore, it is very important to use an adequate amount of acid.

The reason why the egg moves up is the formation of CO_2 from $CaCO_3$ on the egg surface by the action of hydrochloric acid. When

Experiment 17: Magic eggs.

the egg arrives at the surface of the solution the CO_2 bubbles burst, and the egg begins to move downwards.

Waste Disposal   The contents of both cylinders are flushed down the drain.

Reference   L.A. Ford, *Chemical Magic*, Dover Publications, Inc., New York, 1993.

Experiment 18
Colored Kinetics

My ideas of theorizing in the empirical sciences are the following: The one who wants to establish a theory has to prove it with all the facts referring to it, without any prejudice favoring this theory, he has to stress on its bad and on its good qualities. He never should try to create convictions where there is only probability; for the one who passes probability off as truth will be, if he wants or not, a misleader.

Jöns Jakob Berzelius

Apparatus Three 100-mL beakers, four Pasteur pipettes, 50-mL and 25-mL measuring cylinders, overhead projector, safety glasses, protective gloves.

Chemicals Dan Clorox (cleaning liquid containing sodium hypochlorite), food colors (McCormick) yellow (No. 5E 102), blue (Ponceau 4R E 1214), distilled water. Note: $Na_2S_2O_4$ can be used instead of Dan Clorox.

(i) Green solution: Prepared by mixing 10 drops of each of the blue and yellow solutions. Then, 2 drops of this colored mixture are added to 100 mL of water.
(ii) Bleach solution – two types are prepared:
A: 20 mL of water and 1 drop of Dan Clorox.
B: 20 mL of water and 4 drops of Dan Clorox

Attention! Safety glasses and protective gloves must be used at all times.

Experimental Procedure Two 100-mL beakers are placed on the overhead projector. Each beaker contains 40 mL of the green solution (i). Solutions A and B (ii) are each added to the green solutions. As the colors change from green → yellow → blue, the appropriate times are measured. When using solution B, the color change takes place after about 20 seconds, but with solution A the color changes after about 1 minute.

Explanation It is easy to observe that the solution with the higher concentration of bleach reacts faster, as indicated by the color change of the

Spectacular Chemical Experiments. Herbert W. Roesky
Copyright © 2007 WILEY-VCH Verlag GmbH & Co. KGaA, Weinheim
ISBN: 978-3-527-31865-0

solution, from yellow to blue. The solution of the blue color also reacts with the bleach, but only after some additional time. This experiment shows clearly that the reaction time depends on the concentration of the solutions.

Waste Disposal The solutions can be flushed down the drain.

Reference G.C. Weaver, D.R. Kimbrough, *J. Chem. Educ.* **1996**, *73*, 256.

Experiment 19
Flushing Peppermint Tea

Let rush the molecules,
No matter what's their rules,
Don't slice, don't mystify,
Ecstasies sanctify.

Christian Morgenstern

Apparatus	250-mL beaker (high form), measuring cylinders (20 and 100 mL), spatula, safety glasses, protective gloves.
Chemicals	Ethyl acetate, H_2O_2 35%, oxalic acid-bis(2,4-dinitrophenylester), peppermint tea bag.
Attention!	Hydrogen peroxide can cause burns. Skin and eye contact must be avoided. Oxalic acid and oxalates are toxic and must not be ingested. Safety glasses and protective gloves must be used at all times.
Experimental Procedure	100 mL of ethyl acetate and 20 mL of 35% H_2O_2 are placed in the beaker. Then, one pinch of the oxalic acid-bis-(2,4-dinitrophenylester) is added to the mixture. After addition of a peppermint tea bag the room must be darkened completely. When the beaker is moved, a fiery red light is seen to shine.
Explanation	The oxalic acid ester is oxidatively decomposed by H_2O_2, producing activated carbon dioxide. The chlorophyll in the peppermint tea serves as a photosensitizer. During the relaxation of the activated carbon dioxide, the red fluorescence of the coloring chlorophyll is activated. The ethyl acetate serves to dissolve the chlorophyll.
Waste Disposal	The chemicals are disposed of in the collecting container for organic residues.

Spectacular Chemical Experiments. Herbert W. Roesky
Copyright © 2007 WILEY-VCH Verlag GmbH & Co. KGaA, Weinheim
ISBN: 978-3-527-31865-0

References

– D. Woehrle, M.W. Tausch, W.D. Stohrer, *Photochemie: Konzepte, Methoden, Experimente*, Wiley-VCH, Weinheim.
– Web Site www.theochem.uni-duisburg.de.

Experiment 19: The red fluorescence of chlorophyll.

Experiment 20
Chemiluminescence

Light is the symbol of true serenity.
Light is certainly action –
Light is like life, efficient effect . . .
Light makes fire,
Light is like light.
Every effect is transition. In chemistry both merge into one another . . .

Novalis, *Traktat vom Licht*

Apparatus Beakers: one 5-L, two 2-L, one 400-mL, and one 250-mL. Measuring cylinders: two 100-mL, one 25-mL. Glass rods, safety glasses, protective gloves.

Chemicals Luminol, Na_2CO_3, H_2O_2 (3%), hemin, diluted ammonia solution, rhodamine B, sodium fluorescein, distilled water.

Solutions:

(A): 2 g luminol, 40 g Na_2CO_3, and 70 mL H_2O_2 (3%) are placed in the 400-mL beaker.
(B): 50 mg hemin, 20 mL diluted ammonia solution, and 80 mL H_2O are poured into the 250-mL beaker.

Note: Solutions (A) and (B) should be mixed shortly before being used in the experiment, as they do not store well.

Attention! Safety glasses and protective gloves must be used at all times.

Experimental Procedure First, two-thirds of the solution (B) are added to all of solution (A). Initially, an intense blue light is observed. The blue-shining solution is then poured into the 5-L beaker, which is filled with 4.5 L of water. No color change is observed. A pinch of rhodamine B is placed into one of the 2-L beakers, and a pinch of sodium fluorescein is added to the other 2-L beaker. Then, 1.5 L of the light blue-shining solution

Spectacular Chemical Experiments. Herbert W. Roesky
Copyright © 2007 WILEY-VCH Verlag GmbH & Co. KGaA, Weinheim
ISBN: 978-3-527-31865-0

are poured into each of the 2-L beakers. The solution in the beaker containing rhodamine B will then shine a rose color, whilst the solution in the beaker containing sodium fluorescein turns yellow-green. The chemiluminescence of the three solutions can be intensified by adding H_2O_2 and the remainder of the hemin solution, so that the duration of the chemiluminescence is extended.

Explanation

Electronically activated molecules can change into their ground state by emitting light (fluorescence). If the total spin of the molecule changes, the life of the electronically activated state can increase. The transition into the ground state by the emission of light is slowed down; this normally occurs during transition of the molecule from the activated state into the singlet ground state.

Waste Disposal

Due to their low concentrations, the solution can safely be flushed down the drain.

Reference

R. Brückner, personal communication.

Experiment 20: Chemiluminescense. Left: rhodamine B; center: luminol; right: sodium fluorescein.

Experiment 21
The Colors White-Yellow-Black

It is not easy to outline the text for a text book. This serves another purpose than a manual, in which the strong systematic order is essential. In a text book however a selection has to be made by which science is understood best and kept best in ones mind.

Jöns Jakob Berzelius

Apparatus	A 1-L beaker, magnetic stirrer with a stirring bar, measuring cylinders (500 mL, two 100-mL, 50-mL), safety glasses, protective gloves.
Chemicals	$NaHSO_3$, starch solution, KIO_3, $AgNO_3$, distilled water.
Attention!	KIO_3 is a strong oxidant. Contact with the skin and the eyes must be avoided. Safety glasses and protective gloves must be used at all times.
Experimental Procedure	The following solutions are prepared for the demonstration:

(A): 17 g of $NaHSO_3$ are dissolved in 500 mL of a 5% (w/v) starch solution.
(B): 15 g of KIO_3 are dissolved in 1000 mL of distilled water.
(C): 0.025 M $AgNO_3$ solution (MW $AgNO_3 \cong 169.875$).

Water (300 mL) is poured into the 1-L beaker placed on the magnetic stirrer. Under strong stirring, 50 mL of solution (A), 86 mL of solution (B), and 100 mL of solution (C) are added. The resulting solution turns immediately milky white on color, but after about a further 3–5 seconds the milky white color becomes orange-yellow. Finally, after about 15 seconds, the solution suddenly turns deep blue in color (it actually appears black).

Here, the normally used (but toxic) $HgCl_2$ solution is replaced by an $AgNO_3$ solution.

Explanation The milky white cloudy appearance is due to the presence of $AgIO_3$. During the reduction of $AgIO_3$ with $NaHSO_3$, AgI is then formed, so that the white color turns orange-yellow. Then, the iodate and iodide react in an acidic medium to produce iodine. In the presence of starch, the dark blue (almost black) amylose complex is formed which contains iodine in the form of an I_5^- chain.

Waste Disposal The residues are collected and stored in the container for silver residues.

Reference L.E. Wilkinson, *J. Chem. Educ.* **2004**, *81*, 1474.

Experiment 22
Nitrogen and Hydrogen by Electrolysis

Only a fool does not make experiments.

Charles R. Darwin

Apparatus	One Hofmann electrolysis apparatus, measuring cylinder, one transformer with rectifier, cable, test tubes, wooden splint, two platinum electrodes, safety glasses, protective gloves.
Chemicals	NH_4Cl, NH_4OH solution (4%). Solution (A): Saturated NH_4Cl solution. Solution (B): NH_3 solution (10%) is saturated with NH_4Cl.
Attention!	Hydrogen and air form explosive mixtures. Safety glasses and protective gloves must be used at all times.
Experimental Procedure	A solution is prepared by pouring into the measuring cylinder 10 parts of the saturated NH_4Cl solution (A) and one part of the NH_3 solution (B). The resulting solution is then transferred into the Hofmann electrolysis apparatus. First, electrolysis is carried out for about 25 minutes at 24 V with opened stopcocks, after which both stopcocks at the ends of the electrolysis tubes are closed simultaneously. The experiment is complete after 6–8 minutes. Electrolysis for 25 minutes before closing the stopcocks is necessary in order to saturate the solution with each of the produced gases. Otherwise, the volumes will not represent the exact rates of the amount of formed gases. Saturation in both of tubes is shown by the almost same height of the meniscus of the solutions. At the anode where the nitrogen is formed, a slight yellow color can be seen on top of the solution. This color only appears during saturation of the solution. When the electrolysis is finished, the gas at the anode is collected into a test tube. When this gas is tested, the flame of a burning

Spectacular Chemical Experiments. Herbert W. Roesky
Copyright © 2007 WILEY-VCH Verlag GmbH & Co. KGaA, Weinheim
ISBN: 978-3-527-31865-0

wooden splint is extinguished. At the cathode, the hydrogen is also collected in a test tube; its presence is confirmed by placing the open side of the test tube close to the flame of a Bunsen burner (pop!).

Explanation

At the anode, nitrogen is produced by the oxidation of NH_3. At the cathode, hydrogen is generated by reduction of NH_3. The volumes of nitrogen and hydrogen are in the ratio of 1 to 3. NH_4Cl serves as a conducting salt and as a reagent. Very small amounts of chlorine are formed at the anode, but they cannot be detected by smell.

Waste Disposal

The remaining solution in the Hofmann electrolysis apparatus is diluted with water and then flushed down the drain.

Experiment 23
Demonstration of the Plasma State:
A "sparkling cross"

Experience, not reading and listening, is the thing. It is not a matter of indifference whether an idea enters the soul through the eye or through the ear.

Georg Christoph Lichtenberg

Apparatus	A 1-L beaker, stainless-steel electrode, aluminum electrode formed in the shape of a cross, transformer for voltage up to 200 V direct current, strength of the current more than 10 Amps, safety glasses, protective gloves.
Chemicals	$NaH_2PO_4 \cdot 2\,H_2O$, distilled water.
Attention!	During the operation any contact with the set-up should be avoided. A direct current of 150–200 V may be extremely dangerous. Contact with the current can cause death! Safety glasses and protective gloves must be used at all times.
Experimental Procedure	The sodium dihydrogen phosphate solution ($77\,g\,L^{-1}$) is poured into the beaker; the electrodes which have been attached with a clamp are then immersed into the solution. The cathode is made from stainless steel, and has an area of about $40\,cm^2$ ($6.5 \times 6.5\,cm$). The anode is made from aluminum, and is formed in the shape of a cross; the bars of the cross are each 7 cm long and 1 cm wide. The electrodes are connected to the current supply, the room is darkened, and the direct current of about 150 V is switched on. The anode begins to shine, and "sparkling stars" can be seen. When the current is turned up to 200 V, the sparkling becomes even stronger.
Caution!	The amperage is more than 10 A!
Explanation	Plasma is a hot gas consisting of ions, electrons, and neutral particles. The most important parameters for the description of plasma are

Spectacular Chemical Experiments. Herbert W. Roesky
Copyright © 2007 WILEY-VCH Verlag GmbH & Co. KGaA, Weinheim
ISBN: 978-3-527-31865-0

density, pressure, temperature, and the mean free path between two collisions. Due to the high voltage and high amperage, oxygen atoms are produced at the anode; these oxygen atoms, when recombining to O_2 molecules and transitioning into the ground state, emit light.

Waste Disposal The residues can be flushed down the drain.

Reference J.P. Schreckenbach, K. Rabending, *J. Chem. Educ.* **1996**, *73*, 782.

Part II
The Color Blue

As yellow is always accompanied with light, so it may be said that blue still brings a principle of darkness with it.

This color has a peculiar and almost indescribable effect on the eye. As a color it is powerful. Its appearance, then, is a kind of contradiction between excitement and repose.

As the upper sky and distant mountains appear blue, so a blue surface seems to retire from us.

But as we readily follow an agreeable object that flies from us, so we love to contemplate blue, not because it advances to us, but because it draws us after it.

Blue gives us an impression of cold, and thus, again, reminds us of shade.

Rooms which are hung with pure blue, appear in some degree larger, but at the same time empty and cold.

The appearance of objects seen through a blue glass is gloomy and melancholy.

Johann Wolfgang von Goethe, *Theory of Colors*

Experiment 24
Witching Hour

When the mentally disturbed King Ludwig II of Bavaria had seen the opera *Tannhäuser*, by Richard Wagner, he wished to have all-over blue light – his favorite color – in the "Venus Cave". This led to problems because the most beautiful blue was not good enough. Professor Adolf von Bayer, who later became Nobel laureate, wrote a letter to the industrial chemist Heinrich Caro at BASF to solve the complicated problem:

"... the blue which we used till now is the one which is in the package marked by 'Light Blue'. The king is not content with it; he thinks it is not pure enough. The ideal color for him is the color of Lapis Lazuli. It is certainly not possible to make the cave as dark as Lapis Lazuli and at the same time bright, but it should be possible to grant his wish ...". "... please tell me, if there is a blue more blue than the attached triphenylrosaniline."

Apparatus Glass tube (closed at one side, 60 cm long, ca. 2 cm diameter), funnel, stand, boss, clamp, four 1-L measuring flasks, Bunsen burner, thermometer, three 50-mL measuring cylinders, one 100-mL measuring cylinder, a 250-mL Erlenmeyer flask, safety glasses, protective gloves.

Chemicals H_2O_2 (30%), KSCN, $CuSO_4 \cdot 5H_2O$, NaOH, luminol, distilled water.

Attention! H_2O_2 can cause severe burns; skin and eye contact must be avoided. Copper salts are toxic. Safety glasses and protective gloves must be used at all times.

Experimental Procedure The following solutions are prepared in the four measuring flasks:

Solution (A): (1.0 M H_2O_2) 115 mL of H_2O_2 (30%) are added to 885 mL of water to a final volume of 1 L.
Solution (B): (0.15 M KSCN) 14.55 mL of KSCN solution are added to water to a final volume of 1 L.
Solution (C): (6×10^{-4} M $CuSO_4$) to 0.15 g of $CuSO_4 \cdot 5H_2O$ water is added to a final volume of 1 L.

Spectacular Chemical Experiments. Herbert W. Roesky
Copyright © 2007 WILEY-VCH Verlag GmbH & Co. KGaA, Weinheim
ISBN: 978-3-527-31865-0

Solution (D): (0.10 M NaOH and 3.7×10^{-3} M luminol) 4 g of NaOH are dissolved in 100 mL of water, after which 0.55 g luminol is added under stirring to dissolve it. Water is then added to a final volume of 1 L.

The glass tube is attached to the stand with a boss and a clamp, and the funnel is installed. Now, the solutions are poured into the Erlenmeyer flask: 30 mL of (A), 30 mL of (B), 60 mL of (C), and 30 mL of (D). Without stirring, the flask is heated (using the Bunsen burner) up to 50–60 °C. The mixture is then poured into the glass tube, and the room is completely darkened. Immediately, a pulsing light is emitted from the solution. The luminescence appears about 10 times, after which the experiment comes to an end.

Explanation

By adding luminol, an oscillating luminescence is observed. It is important to use the solutions (A) to (D) in a ratio of 1:1:2:1. It is important NOT to use NH_4SCN and NH_4OH instead of KSCN and NaOH, as the NH_4^+ ion hinders the oscillations. Spectators are always fascinated by light-emitting reactions!

Waste Disposal

The solutions are concentrated on the water bath, and the residues disposed of in the container for collecting heavy metals.

Reference

H.E. Prypsztejn, *J. Chem. Educ.* **2005**, *82*, 53–54.

Experiment 25
Molybdenum Blue

Mind is able to repeat what has been,
What nature builds, he builds it, choosing, too.
Far away from nature builds reason, but only into the unsubstantial,
You only, genius, promote nature according to nature.

Friedrich Schiller, *Aphorismen*

Apparatus	A 100-mL beaker, safety glasses, protective gloves.
Chemicals	$(NH_4)_6Mo_7O_{24} \cdot 4H_2O$, 2 M HCl, distilled water, granulated zinc.
Attention!	Safety glasses and protective gloves must be used at all times.
Experimental Procedure	An ammonium molybdate solution (3%) (ca. 25 mL) is acidified with 2 M hydrochloric acid and added to several pieces of zinc. The solution turns into a deep blue color.
Explanation	Although the molybdenum blue solutions have been known for over 200 years, their structures have only been resolved during the past few years by A. Müller and co-workers. It is interesting to know that wheel-shaped polyoxomolybdates of composition $\{Mo_{11}\}_{14}$ and $\{Mo_{11}\}_{16}$ have been found. These isolated macro ionic species have diameters in the range of 2.5 to 6 nm, and they may contain up to 368 molybdenum atoms. These supramolecular clusters demonstrate a completely new chemistry.
Waste Disposal	The heavy metal-containing solutions are disposed of in the container for collecting heavy metals.
References	– A. Müller, P. Kögerler, A. Dress, *Coord. Rev.* **2001**, *222*, 193. – Müller, S. Roy, *Coord. Rev.* **2003**, *245*, 153. – T. Liu, E. Diemann, H. Li, A. Dress, A. Müller, *Nature* **2003**, *426*, 59.

Spectacular Chemical Experiments. Herbert W. Roesky
Copyright © 2007 WILEY-VCH Verlag GmbH & Co. KGaA, Weinheim
ISBN: 978-3-527-31865-0

Experiment 26
Combustion of Sulfur in Oxygen

Reports of the use of sulfur date back to 400 BC, when the Greek historian Thukydides described how the Boeotians used burning sulfur when they attacked a village.

Apparatus A 2-L flat-bottomed flask, Bunsen burner, combustion boat, safety glasses, protecting gloves.

Chemicals Sulfur flowers, O_2, litmus solution, distilled water.

Attention! Sulfur dioxide is a severe respiratory irritant. Therefore, this reaction should be carried out in a well-ventilated fume hood. Safety glasses and protective gloves must be used at all times.

Experimental Procedure Approximately 200–300 mL of water colored with litmus (blue to purple) are poured into the flat-bottomed flask. After filling the flask with oxygen (from a gas cylinder), the combustion boat, which is filled with sulfur flowers, is heated with the Bunsen burner. The sulfur melts and begins to burn with a weak blue flame. When the boat is placed into the flask, the sulfur burns much more strongly, with a light blue flame producing a white fog.

As soon as the flame is extinguished the flask is strongly agitated so that the resulting gas is dissolved in the water. The litmus solution, which was blue at the start of the experiment, then turns red, providing typical proof of an acid reaction.

Explanation Elemental sulfur burns in air or in pure oxygen yielding sulfur dioxide. This reacts with water, yielding protons and hydrogen sulfite.

$$SO_2 + 2\,H_2O \rightarrow H_3O^+ + HSO_3^-$$

In the case of a higher SO_2 concentration, hydrogen disulfite might be formed.

Waste Disposal The slightly acidic solution can be flushed down the drain.

Spectacular Chemical Experiments. Herbert W. Roesky
Copyright © 2007 WILEY-VCH Verlag GmbH & Co. KGaA, Weinheim
ISBN: 978-3-527-31865-0

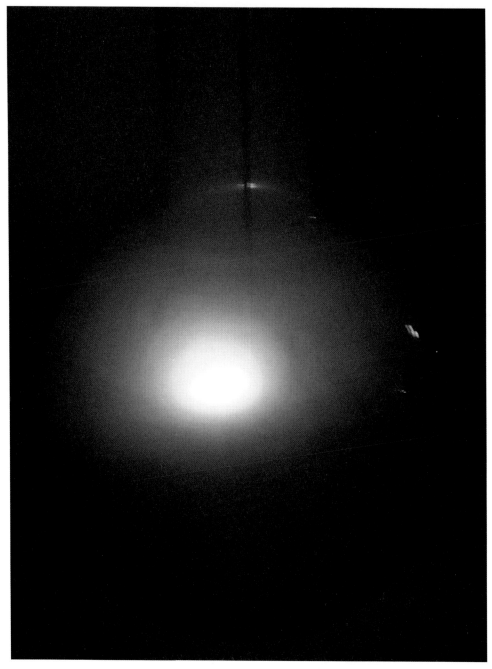

Experiment 26: The combustion of sulfur in oxygen.

Experiment 27
Phosphorus Salt Pearl or Cobalt Salt Pearl

It is the secret of every inventive genius to believe that nothing is impossible.

Justus von Liebig

Apparatus	Two small watch-glasses, magnesia sticks, Bunsen burner, safety glasses, protective gloves.
Chemicals	$NaNH_4HPO_4$, $Co(NO_3)_2$.
Attention!	Safety glasses and protective gloves must be used at all times.
Experimental Procedure	A pinch of the above-mentioned chemicals is placed on each of the watch glasses. The magnesia stick is then placed at one end into the flame of the burner and heated to red heat, and then quickly dropped into $NaNH_4HPO_4$ and again heated with the burner. It can be observed that the melt is degassing. When the melt is clear, it is briefly brought into contact with the cobalt nitrate, and then heated again. The melt turns blue, and the metal oxide is dissolved by the sodium phosphate.
Explanation	The cobalt oxide is formed by heating $Co(NO_3)_2$, and dissolves in the sodium phosphate melt in the form of a pearl to yield blue $NaCoPO_4$.
Waste Disposal	The residues of the phosphoric salt pearl are disposed of in the normal garbage.

Spectacular Chemical Experiments. Herbert W. Roesky
Copyright © 2007 WILEY-VCH Verlag GmbH & Co. KGaA, Weinheim
ISBN: 978-3-527-31865-0

Experiment 27 Phosphorus Salt Pearl or Cobalt Salt Pearl

Experiment 27: Cobalt salt pearl on a magnesia stick.

Experiment 28
Fehling's Solution

Anyhow, it is typical for the progress that is looks more important than it really is.

Johann Nestroy (1801–1867)

Apparatus	A 100-mL beaker, 10-mL and 20-mL measuring cylinders, glass rod, tripod, Bunsen burner, wire gauze, safety glasses, protective gloves.
Chemicals	$CuSO_4 \cdot 5H_2O$, tartaric acid, NaOH, D(+) glucose.
Attention!	Copper salts are toxic. Contact with the skin should be avoided. Safety glasses and protective gloves must be used at all times.
Experimental Procedure	In the beaker, ca. 10 mL of 0.1 M $CuSO_4$ solution are added to ca. 20 mL of tartaric acid (20% w/v). After adding a few mL of 10 M sodium hydroxide, the solution becomes deep blue. When the deep blue solution is warmed with a few mL of the glucose solution, red copper(I) oxide deposits after a short time.
Explanation	Fehling's solution is prepared by adding copper sulfate to the alkaline solution of a salt of tartaric acid (HOOCCHOHCHOHCOOH). A deep-blue solution is formed which contains the complex Cu(II) ion. Fehling's solution serves as a reagent for aldehydes. By reacting with an aldehyde, copper is reduced to the monovalent state, and a precipitate of copper(I)oxide is formed.

$$RCHO + 2\,Cu^{2+} + NaOH \xrightarrow{H_2O} RCOONa + Cu_2O + 4\,H_3O^+$$

Waste Disposal	The residues are collected and transferred into the container for heavy metals.

Experiment 28 Fehling's Solution

Experiment 28: Fehling's solution. Left: copper sulfate solution; center: after the addition of tartaric acid; right: copper(I) oxide.

Experiment 29
Activated Carbon Decolorizes Water Blue

Everything is poisonous, only the dose determines that something is not poisonous.
Paracelsus

Apparatus	Test tube (20 × 2 cm) with a stopper (or a cylinder), funnel, folded filter, beaker, safety glasses, protective gloves.
Chemicals	Granulated activated carbon, water blue solution.
Attention!	Safety glasses and protective gloves must be used at all times.
Experimental Procedure	In the test tube, granulated activated carbon is added to ca. 20 g of water blue, and then stirred vigorously. After filtration the filtrate becomes colorless. The carbon has adsorbed the coloring substance. Water blue: other names are cotton blue, acid blue 93, Porrier's blue, aniline blue water-soluble.
Explanation	Activated carbons are porous carbons with a large surface area. They serve as adsorbents for coloring substances and other organic substances, such as solvents. Toxic substances in the air may also be adsorbed by activated carbon (e.g., filters for gasmasks).
Waste Disposal	The adsorbed activated carbon is placed into the container used for collecting solid inorganic substances.

Experiment 30
Blue Bottle – The Blue Miracle

And then take some trouble not to write such nonsense, facts, facts, and facts, and the most important is to make it short.

Fjodor M. Dostojewski

Apparatus	One 500-mL Erlenmeyer flask with a stopper, 1-mL pipette, 250-mL measuring cylinder, safety glasses, protective gloves.
Chemicals	D(+) glucose, dextrose, NaOH, methylene blue 0.2%, distilled water.
Attention!	Safety glasses and protective gloves must be used at all times.
Experimental Procedure	45 g of D(+) glucose dissolved in 250 mL of H_2O, and 1 mL of 0.2% methylene blue solution, are poured sequentially into the Erlenmeyer flask. The resultant solution becomes colorless after a short time. Strong shaking makes the solution turn blue again, but after a while it becomes colorless again. This color change can be repeated again and again.
Explanation	All eventual stages of the redox reactions appear in the formula. It is possible that some of the partial reactions occur simultaneously.
Waste Disposal	The solution from the Erlenmeyer flask and the organic residues are collected and disposed of.
References	– J.A. Campbell, *J. Chem. Educ.* **1963**, *40*, 578. – J.A. Campbell, *Why do chemical reactions occur?* Prentice-Hall, 1965.

Experiment 30 Blue Bottle – The Blue Miracle

glucose + H$_2$O → (Oxidation) → gluconic acid + 2H

methylene blue (blue) [Cl$^-$ + 2H] → (reduction) → methylene blue (colorless) + HCl

H$_2$O ← 1/2 O$_2$

Experiment 31
Generation of Blue (N$_2$O$_3$) Dinitrogen Trioxide

Art and science is not all we need, this business requires patience too.

Johann Wolfgang von Goethe

Apparatus	Test tube (20 × 2 cm), stand, boss, clamp, Dewar vessel (or beaker), safety glasses, protective gloves.
Chemicals	NaNO$_2$, methyl alcohol (or liquid nitrogen), solid CO$_2$, H$_2$SO$_4$ 8% (18 M) (or wet ice salt mixture).
Attention!	NO and NO$_2$ gases are highly toxic. This experiment must be carried out in a well-ventilated fume hood. Safety glasses and protective gloves must be used at all times.
Experimental Procedure	A pinch of solid sodium nitrite is placed in the test tube. After adding about 1–2 mL of sulfuric acid, brown NO$_2$ gas is formed after a short time. The gas-containing test tube is immersed into the cold bath of solid carbon dioxide and methyl alcohol in the Dewar vessel. After a short period of cooling, blue N$_2$O$_3$ is formed in the test tube. In a wet ice salt mixture with a temperature of −15 °C, the blue N$_2$O$_3$ can be condensed. However, it is much more clearly seen when lower temperatures are used.
Explanation	N$_2$O$_3$ consists of NO and NO$_2$. The sulfuric acid reacts with NaNO$_2$, producing HNO$_2$. The nitrous acid (HNO$_2$) decomposes under elimination of water according to the following equation: $2\ HNO_2 \rightarrow NO + NO_2 + H_2O$

Experiment 32
Bleaching with a Household Product

Practice is much more imaginative than theory.

Manfred Rommel

Apparatus	Three 100-mL beakers (or glass dishes), Pasteur pipettes, 25-mL and 50-mL measuring cylinders, overhead projector, safety glasses, protective gloves.
Chemicals	Dan Clorox (cleaning liquid containing sodium hypochlorite), food colors (McCormick) yellow (No. 5E 102), blue (Ponceau 4R E 124), distilled water.
Attention!	Safety glasses and protective gloves must be used at all times.
Experimental Procedure	10 drops of the blue and, respectively, the yellow food color, produce the green solution. For the demonstration, two drops of this solution are added to 100 mL of H_2O. Two different concentrations of bleach solution are prepared:

  (i) 20 mL of H_2O and 1 drop of Dan Clorox (bleach solution).
  (ii) 20 mL of H_2O and 4 drops of Dan Clorox (bleach solution)

The three beakers are placed on the overhead projector and each of them is filled with 40 mL of the green solution. Solution (i) is placed into the first beaker, and solution (ii) into the second beaker. (The solution in the third beaker serves for comparison).

After about 1 minute, the color in the first beaker changes from green-yellow to green-blue-green and finally to light blue. In the second beaker, the same color changes occur, but after about 20 minutes.

As a variant of this experiment, a Petri dish (10 cm diameter) is filled with the green solution (four drops of green-colored solution are added to 50 mL of H_2O) and placed on the overhead projector.

Spectacular Chemical Experiments. Herbert W. Roesky
Copyright © 2007 WILEY-VCH Verlag GmbH & Co. KGaA, Weinheim
ISBN: 978-3-527-31865-0

After adding 20 drops of the undiluted bleach solution, the blue coloring penetrates the green solution. (When 20 drops of the bleach solution are added under stirring, a rapid color change from green to blue occurs).

Explanation

The NaOCl dissociates in water into OCl^-_{aq}, Na^+_{aq}, and hydroxide. The hypochlorite decomposes into chloride and atomic oxygen, and the latter bleaches the color.

Waste Disposal

The solutions can be poured down the drain.

References

– G.C. Waever, D.R. Kimbrough, *J. Chem. Educ.* **1996**, *73*, 256.
– J.J. Fortman, *J. Chem. Educ.* **1994**, *71*, 848.

Experiment 33
Ink Blue – Solvated Electrons

In travelling over the Harz in winter, I happened to descend from the Brocken towards evening; the wide slopes extending above and below me, the heath, every insulated tree and projecting rock, and all masses of both, were covered with snow or hoar-frost. The sun was sinking towards the Oder ponds. During the day, owing to the yellowish hue of the snow, shadows tending to violet had already been observable; these might now be pronounced to be decidedly blue, as the illuminated parts exhibited a yellow deepening to orange. But as the sun at last was about to set, and its rays, greatly mitigated by the thicker vapors, began to diffuse a most beautiful red color changed to a green in beauty to the green of the emerald.

Johann Wolfgang von Goethe, *Theory of Colors*

Science is made by men.

Werner Heisenberg

Apparatus A 50-mL trap with inlet and delivery tubes and a dry ice condenser, stand, clamps, bosses, PVC tubes, knife, safety glasses, protective gloves.

Chemicals Sodium, pressure cylinder with ammonia, dry ice (solid CO_2), acetone.

Attention! An ammonia solution containing dissolved alkali metals slowly decomposes to form alkali metal amides and hydrogen. Ammonia is a pungent, irritating gas. Therefore, the experiment must be carried out in a well-ventilated fume hood. Safety glasses and protective gloves must be used at all times.

Experimental Procedure The trap with the dry ice condenser is vertically attached to the stand with clamps. It is then connected with the PVC tube to the pressure cylinder with ammonia. The condenser is filled with dry ice. For better thermal exchange, acetone is added to the dry ice. Next, a strong ammonia stream is passed into the trap, whereupon drops form very rapidly at the cold finger and the cold bath releases large

Spectacular Chemical Experiments. Herbert W. Roesky
Copyright © 2007 WILEY-VCH Verlag GmbH & Co. KGaA, Weinheim
ISBN: 978-3-527-31865-0

amounts of CO_2. When 3–5 mL of liquid ammonia have condensed inside the trap, the ammonia stream is disconnected. The cold finger is removed and a pea-sized piece of sodium is added. Immediately afterwards the colorless strong refracting liquid ammonia turns into an intensive blue color.

Explanation

In liquid ammonia, sodium and other alkali metals dissolve under the formation of solvated cations $[Na(NH_3)_x]^+$ and solvated electrons $[e(NH_3)_y]^-$. Due to their excitation by light in the range of 1500 nm, the solvated electrons produce the blue solution color. The addition of a rusty iron wire causes the blue color to be destroyed. Iron(II) compounds react catalytically under the formation of hydrogen and $NaNH_2$.

Waste Disposal

The residues in the trap are decomposed by alcohol and, after neutralization, may be flushed down the drain.

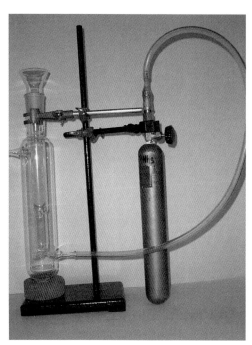

Experiment 33: The experimental set-up before the condensation of the NH_3 (left) and after the addition of metallic sodium (right).

Part III
The Color Red

We are here to forget everything that borders on yellow or blue. We are to imagine an absolutely pure red, like fine carmine suffered to dry on white porcelain. We have called this color "purple" by way of distinction, although we are quite aware that the purple of the ancients inclined more to blue.
The effect of this color is as peculiar as its nature. It conveys an impression of gravity and dignity, and at the same time of grace and attractiveness. The first in its dark deep state, the latter in its light attenuated tint; and thus the dignity of age and the amiableness of youth may adorn itself with degrees of the same hue.

Johann Wolfgang von Goethe, *Theory of Colors*

Experiment 34
Purple or Colorless: An Entertaining Demonstration

I think that the ways to knowledge are as admirable as the nature of the things itself.

Johannes Kepler

Apparatus	Four 50-mL beakers, one 100-mL measuring cylinder (or stand cylinder), safety glasses, protective gloves.
Chemicals	n-hexane, $KMnO_4$, iodine, distilled water.
Attention!	$KMnO_4$ is a strong oxidant. Safety glasses and protective gloves must be used at all times.
Experimental Procedure	For the demonstration, four solutions are prepared as follows:

1. The first beaker contains 40 mL of n-hexane.
2. The second beaker contains 40 mL of a $KMnO_4$ solution acidified with sulfuric acid, which must have the same color as the solution in beaker 4.
3. The third beaker contains 40 mL of water.
4. The fourth beaker contains 40 mL of a solution of iodine in n-hexane (the same color as that in beaker 2).

The solutions of beakers 1 and 2 are poured into the first measuring cylinder. Two phases are formed – the lower phase is colored, and the upper one is colorless. When the solutions of beakers 3 and 4 are transferred into the second measuring cylinder, the phases are inverted – the lower phase is colorless, and the upper one is purple.

Explanation   It is important that the $KMnO_4$ solution is acidified. In this solution the MnO_4^- is reduced to manganese(II). Otherwise, the permanganate is reduced under elimination of oxygen according to the following equation:

Spectacular Chemical Experiments. Herbert W. Roesky
Copyright © 2007 WILEY-VCH Verlag GmbH & Co. KGaA, Weinheim
ISBN: 978-3-527-31865-0

$$4 \text{ MnO}_4^- + 2 \text{ H}_2\text{O} \rightarrow 4 \text{ MnO}_2 + 3 \text{ O}_2 + 4 \text{ OH}^-$$

The yellow-brown color of iodine in solvents such as water, alcohols or amines is due to the charge-transfer from the solvent to iodine. The permanganate color is also due to a charge-transfer from oxygen to manganese(VII).

Waste Disposal Hexane and hexane iodine mixtures must be collected and disposed of completely. The aqueous permanganate solution can be reduced with H_2O_2 and poured down the drain.

Reference T.M. Kitson, *J. Chem. Educ.* **2003**, *80*, 892.

Experiment 35
A "Red Component" in Newspapers

Conversation of the elements
but the boron, but in their well
the essential oils; who asks zinc and cyanide,
who cares about the colloids, the hatred
between calcium and arsenic, the radicals' love
of water, the transuranian silent obsession?

nobody reads the manifests of the rare earths,
the secret of the salts, sealed inside druses,
is unsolved, nobody sings of the old fight,
between left handed and right handed aldehydes,
unmentioned the gossip of the hormones, arrogance
pushes the crystals, the silicates
talk about gravel, the spars, the blends
whisper, the oxalic acids and asbestos; the ether
in its ampoules agitates against the sulfur, the iodine
and the glycerine. full of hostility
lead sugar phosphorus and sublimate are waiting in blue bottles.
murderers!
messengers! helpless witnesses of the world!
why can I not counter and extinguish fire,
uninvite guests, cancel the delivery of milk and newspaper,
immerge into the soft talk of the resins,
the bases, the minerals, to go deep into
the endless mediating and moaning of the elements, to remain
in the silent monologue of the substances?

Hans Magnus Enzensberger

Apparatus	Brush, 50-mL beaker, safety glasses, protective gloves.
Chemicals	1,2,5-Trihydroxybenzene, 12 M hydrochloric acid (37%), distilled water, wood-free writing or drawing paper, filter paper, newspaper, advertising flyers or other printed material.
Attention!	1,2,5-Trihydroxybenzene is a toxic material. Safety glasses and protective gloves must be used at all times.

Spectacular Chemical Experiments. Herbert W. Roesky
Copyright © 2007 WILEY-VCH Verlag GmbH & Co. KGaA, Weinheim
ISBN: 978-3-527-31865-0

Experiment 35 A 'Red Component" in Newspapers

Experimental Procedure

0.1 g of 1,2,5-trihydroxybenzene are dissolved in 8 mL (12 M) of hydrochloric acid and diluted with 8 mL of water. This solution serves as a test reagent for lignin. Lignin reacts very clearly with hydrochloric acid and 1,2,5-trihydroxybenzene, producing a red color.

Explanation

Wood-free paper, filter paper, etc., becomes yellowish in the presence of a hydrochloric 1,2,5-trihydroxybenzene solution. Conventional and cheap paper containing lignin becomes immediately deep red. In such a way one can show that even serious printed matters may contain a color-sensitive component. In this experiment, even "conservative" newspapers show a red component!

Waste Disposal

The remaining 1,2,5-trihydroxybenzene is transferred into the container used for collecting organic residues.

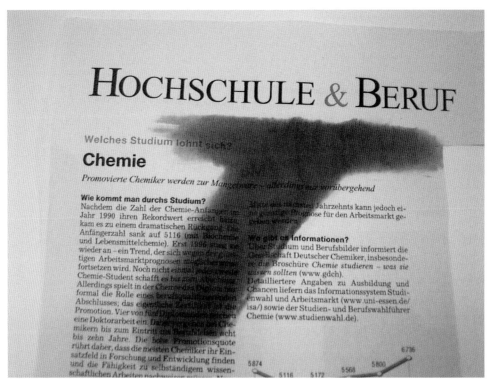

Experiment 35: The red component in newspapers.

Experiment 36
Bleaching of Tomato Juice with Chlorine on a Micro Scale

To catch the phenomena, to fix them with experiments, to arrange the experiences and to know how they will work, with the first to become very observant, with the second as precise as possible, with the third complete, and with the forth to stay broad minded, this means to work through your poor self, of which I never had a precise idea.

Johann Wolfgang von Goethe, Letter to Jacobi, December 29, 1794

Apparatus	Stand, bosses, clamps (or special support), 5-mL syringe, 4-mL glass cuvette, 4-mL glass bottle with a delivery tube and a septum, spatula, Pasteur pipette, safety glasses, protective gloves.
Chemicals	$KMnO_4$, 6 M HCl, tomato juice.
Attention!	Chlorine is highly toxic! The experiment should be carried out in a well-ventilated fume hood. $KMnO_4$ is a strong oxidant. Contact with organic materials should be avoided. Safety glasses and protective gloves must be used at all times.
Experimental Procedure	The glass cuvette is half-filled with tomato juice. About 0.5 g of $KMnO_4$ are placed in the glass bottle, which is closed with a septum connected to the delivery tube and fixed on the stand (or another support). The other end of the delivery tube is inserted into the cuvette with the tomato juice. The syringe which is placed above the small glass bottle (contents ca. 1–2 mL 6 M HCl) is lowered, and the needle of the syringe enters through the septum into the inside of the bottle. A few drops of the acid are added to the $KMnO_4$. Chlorine is formed and passes through the tomato juice. The red juice is bleached after a very short time (1 minute).
Explanation	In an aqueous solution, chlorine reacts yielding HOCl and HCl. During the bleaching reaction HOCl decomposes into HCl and

Spectacular Chemical Experiments. Herbert W. Roesky
Copyright © 2007 WILEY-VCH Verlag GmbH & Co. KGaA, Weinheim
ISBN: 978-3-527-31865-0

oxygen; the latter functions as an oxidizing agent, and consequently bleaches the tomato juice.

Waste Disposal The residues of the permanganate solution are disposed of in the container for collecting heavy metals.

Experiment 37
Production of Non-Drinkable Red Wine

Sometimes one has to drink to make the ideas and the wrinkles in one's brain a little bit smoother, and to recreate the former wrinkles.

Georg Christoph Lichtenberg

Apparatus Two 800-mL beakers, 0.7-L wine bottle, wine glass, funnel, three 5-mL Pasteur pipettes, 1-L measuring cylinder, safety glasses, protective gloves.

Chemicals Na_2CO_3 (soda), $FeCl_3$, NH_4SCN.

Attention! Safety glasses and protective gloves must be used at all times.

Experimental Procedure Before starting the demonstration, both of the beakers and the wine bottle are specially treated. 3 mL of the Na_2CO_3 solution (5%) and 700 mL of H_2O are poured into beaker 1. Beaker 2 contains 2 mL of $FeCl_3$ solution (30%), while 2 mL of the NH_4SCN solution (30%) is placed into the wine bottle.

During the demonstration, the contents of beaker 1 are added to those of beaker 2; the resultant solution is of a light yellow color. This solution is poured into the wine bottle in order to produce "fermentation in the bottle". When the wine is poured into the wine glass it resembles real red wine.

Explanation The hexa-aqua cations of the iron(III) react with the SCN⁻ anions to produce deep red iron(III)thiocyanate complexes. The formation of one such complex is shown by the following equation:

$$[Fe(H_2O)_6]^{3+} + 3\ SCN^- \rightarrow [Fe(SCN)_3(H_2O)_3] + 3\ H_2O$$
light yellow   colorless   deep red

The complexes which contain less thiocyanate anions show a light red color. Moreover, further anionic complexes can be formed by

Experiment 37 Production of Non-Drinkable Red Wine

displacing water molecules in $[Fe(SCN)_3(H_2O)_3]$, when the concentration of the solution is increased by thiocyanate anions.

Waste Disposal The solutions can be flushed down the drain.

Experiment 37: The production of red wine. The beaker in the middle contains water, and the one on the left contains iron(III) chloride solution. In the wine bottle is the NH_4SCN solution and the $FeCl_3$ solution.

Experiment 38
Red Wine as a Color Indicator

Even in the old days, people appreciated a good Burgundy wine. Napoleon is supposed to have lost the battle of Waterloo, because he did not have his daily bottle of Chambertin.

When the Bishop reproached Erasmus von Rotterdam for drinking Musigny during Lent, he answered: "Only my heart is catholic, Your Eminence, my stomach is protestant."

The Bishop of Lyon bought the most expensive wine of the Côte d'Or to use as Communion wine. When reproached by the Cardinal, the Bishop retorted that he did not want to have to pull a face when looking up to God.

Apparatus

Two wine glasses (or 250-mL beakers), glass rods, safety glasses, protective gloves.

Chemicals

Red wine (or rosé wine), 2 M NaOH, 1 M H_2SO_4.

Attention!

After having added sodium hydroxide to the red wine, it SHOULD NOT BE CONSUMED! Danger of cauterization! Safety glasses and protective gloves must be used at all times.

Experimental Procedure

About 100 mL of red wine are poured into each of the wine glasses. When sodium hydroxide is added to the first glass, the wine turns dark green (it looks like used motor oil). After adding sulfuric acid, the dark green color turns into the typical color of red wine. This color change can be repeated endlessly by adding alternately sodium hydroxide and sulfuric acid.

The wine in the second wine glass is used for comparison. When rosé wine is used instead of red wine, the color does not change as much after adding sodium hydroxide. The green is somewhat lighter, which is better seen by the spectator.

Explanation

A similar experiment was first realized by Mariotte, who used sulfur dioxide for acidification. As with acid–base indicators, the color of the dyestuff changes through the protolysis:

Spectacular Chemical Experiments. Herbert W. Roesky
Copyright © 2007 WILEY-VCH Verlag GmbH & Co. KGaA, Weinheim
ISBN: 978-3-527-31865-0

$$\text{HInd} + \text{H}_2\text{O} \rightleftharpoons \text{H}_3\text{O}^+ + \text{Ind}^-$$
indicator acid　　　　　　conjugated indicator base

The indicator acid and indicator base have different colors.

Waste Disposal　　　The solutions can be flushed down the drain.

Part IV
Colloids, Sols, and Gels

The role of the infinite small seems to be infinitely big.

Louis Pasteur

In 1914, Wolfgang Oswald published his book entitled: *The universe of the neglected dimensions*, in which he says: "I do not know any branch of present sciences where as many different and various fields of interest are touched as in colloid chemistry. Of course, also atomic theory and radioactivity are actually exciting for every intellectually interested person. But these are spiritual delicacies compared with colloid chemistry, which nowadays is as necessary as our daily bread for many theoretical and practical fields."

In 1923, Richard Zsigmondy (Nobel Prize 1925) writes the following remarks in the foreword of the series "Colloid research in separated representations" which was edited by himself:

"... Due to the involvement of a big number of researchers having a very different educational background the demands on behalf of the exact sciences on scientific work very often could not be met; and as the number of often inconsistent information was increasing due to the rapid production, it is not surprising that the colloid chemistry did not acquire good reputation for the accurately working researchers but has even lost credit; there are many reasons: in the beginning of this century the tendency to brush a subject away with some commonplaces, the preference of the deductive, philosophical treatment of a subject, ignoring the chemical character of the considered systems have contributed to a large degree to the generalization of detailed observations."

Colloids are particles which range in size from 1 nm to 1000 nm. Therefore, they cannot be recognized without a microscope. They exist in natural water and they have a large surface area. For example, 1 g of sand with a particle diameter of 1 mm has a surface of 30 cm^2. The same amount with a particle size of 10 nm has a surface of 30 m^2.

Experiment 39
Silica Gel from Alkali Silicates

The pleasure of observing and understanding is the most beautiful gift.

Albert Einstein

Apparatus	A 250-mL beaker or goblet, one glass rod, safety glasses, protective gloves.
Chemicals	Sodium silicate solution, distilled water, 1 M H_2SO_4.
Attention!	Safety glasses and protective gloves must be used at all times.
Experimental Procedure	The purchased sodium silicate solution is diluted with distilled water in a ratio of 1:1. About 50 mL of 1 M H_2SO_4 are added to 100 mL of this solution, whereupon a solid gel precipitates. Sometimes the gel can be removed from the beaker in form of a "block".
	When, instead of sulfuric acid, concentrated hydrochloric acid is added to the 1:1 solution, two phases result – a solid phase and a liquid one. This procedure is not recommended.
	If the concentrated hydrochloric acid is added under stirring to the undiluted silicate solution, a solid gel precipitates, like the one already described above.
Explanation	Alkali silicates have the composition of $M_2O \cdot nSiO_2$ (M = Na, K). Sodium silicate has often the composition of $Na_2Si_3O_7$. These are solids which are water-soluble. Acidification yields silicic acids ($H_2Si_3O_7$) which, under the elimination of water, form polysilicic acids and finally condense to a silica-hydrogel.
Waste Disposal	The solid gel can be flushed down the drain.

Experiment 40
Red Gold

Our science is sensual.

Ralph Waldo Emerson

Plan the difficult as long as it is easy, do the great as long as it is small. All the hard work on earth in the beginning is easy. Everything which is great here on earth always starts small.

Laotse

Apparatus	A 400-mL beaker, 5-mL Pasteur pipette, glass rod, tripod, wire gauze, Bunsen burner, thermometer, safety glasses, protective gloves.
Chemicals	$HAuCl_4 \cdot xH_2O$, tannin, distilled water. A solution of 1 g of $HAuCl_4 \cdot xH_2O$ in 500 mL of distilled water is used for the experiment.
Attention!	Safety glasses and protective gloves must be used at all times.
Experimental Procedure	Using a 400-mL beaker, 200 mL of water are heated to about 70 °C, after which 2.5 mL of a $HAuCl_4 \cdot xH_2O$ solution (0.2%) are added via the Pasteur pipette. To this almost colorless solution a freshly prepared tannin solution (0.1%) is added dropwise. The solution turns to an intense red color. The reduction produces the red gold sol.
Explanation	The different colors of the gold colloids depend on the size of the particles.
Waste Disposal	The gold solutions are disposed of separately into the container for collecting gold waste.

Experiment 41
Red Gold Sol

To make gold would be glorious and fine, if all the trouble was not in vain.

German proverb

Apparatus	A 250-mL beaker, glass rod, 5-mL measuring cylinder, 4-mL pipette, Pasteur pipette, tripod, wire gauze, Bunsen burner, safety glasses, protective gloves.
Chemicals	$HAuCl_4 \cdot xH_2O$ (yellow), K_2CO_3, formaldehyde solution (37%), distilled water.
Attention!	Formaldehyde solutions are toxic. Safety glasses and protective gloves must be used at all times.
Experimental Procedure	For the demonstration, the following solutions are prepared:

Solution (A): 1 g of gold chloride (yellow) is dissolved in 50 mL of distilled water.
Solution (B): 0.05 M potassium carbonate solution is prepared.

2 mL of solution (A) are diluted with 100 mL of distilled water, after which 4 mL of the K_2CO_3 solution (B) are added. The resulting solution is heated until boiling. After having added 3 drops of the formaldehyde solution (37%), the resulting solution is boiled for 15–20 seconds. The Bunsen burner is turned off. After about 1 minute the solution turns to a rose color, and after 2 minutes it turns to red.
If 5 drops of formaldehyde are added, the solution turns red more rapidly and more intensely, but it soon becomes cloudy.

Explanation	Formaldehyde reduces Au(III) ions to metallic gold.
Waste Disposal	The gold residues are collected separately for recycling.

Spectacular Chemical Experiments. Herbert W. Roesky
Copyright © 2007 WILEY-VCH Verlag GmbH & Co. KGaA, Weinheim
ISBN: 978-3-527-31865-0

Experiment 42
Blue Gold Sol

Six things are imperative for making gold:
To work hard day and night,
To fan the fire permanently,
To feel smoke and vapor,
To be infected alone,
To loose face and health,
And finally, to feel sadly the deceit.

Anonymous proverb

Apparatus A 250-mL beaker, glass rod, 2-mL pipette, 250-mL measuring cylinder, 10-mL measuring cylinder, tripod, wire gauze, Bunsen burner, safety glasses, protective gloves.

Chemicals $HAuCl_4 \cdot xH_2O$ (yellow), ethyl alcohol, distilled water.

Attention! Safety glasses and protective gloves must be used at all times.

Experimental Procedure For this demonstration, 1 g of $HAuCl_4 \cdot xH_2O$ is dissolved in 500 mL of distilled water. 2 mL of the gold solution, 150 mL of distilled water, and 10 mL of ethyl alcohol are poured into the beaker. After some minutes a blue gold sol results which later turns red. A higher concentration of the gold solution (about 3–4 mL) makes the solution become light blue, then dark blue, and finally purple. The solution can be kept for a longer time.

Explanation The gold colloids range from 210 to 100 nm in size. The solution seems to be clear, although a light beam can be traced when passing through it. Light is scattered at the nanoparticles so that a luminous turbidity can be observed. This effect was first examined in 1857 by Michael Faraday and the English physicist John Tyndall. This effect is known as the "Tyndall effect" in the latter's honor.

Waste Disposal The gold residues are collected separately for recycling.

Spectacular Chemical Experiments. Herbert W. Roesky
Copyright © 2007 WILEY-VCH Verlag GmbH & Co. KGaA, Weinheim
ISBN: 978-3-527-31865-0

Experiment 42 Blue Gold Sol

Experiment 42/41: Left: red gold sol. Right: blue gold sol.

Experiment 43
Cherry Red Gold Sol

"El Dorado" is the legendary country of the Golden Man which the Spanish Conquistadores thought to be in South America. The legend has probably its origin in the habit of the Chibcha Indians in Columbia. Every year they elected a new chief whom they covered completely with gold. Then they washed him in a consecrated lake, and threw more gold and precious stones into the water.

Apparatus A 250-mL beaker, glass rod, 2-mL and 5-mL Pasteur pipettes, tripod, wire gauze, Bunsen burner, 100-mL measuring cylinder, safety glasses, protective gloves.

Chemicals $HAuCl_4 \cdot xH_2O$ (yellow), K_2CO_3, D(+)glucose, distilled water.

Attention! Safety glasses and protective gloves must be used at all times.

Experimental Procedure For this demonstration the following solutions are prepared:

Solution (A): $HAuCl_4$ solution (1%).
Solution (B): K_2CO_3 solution (0.15%).
Solution (C): D(+)glucose solution (0.5%).

100 mL of distilled water, 2 mL of gold solution (A) (1%), and 0.5 mL of the K_2CO_3 solution (B) (0.05 M) are poured into the beaker on the wire gauze and heated until boiling. Subsequently, 1 mL of solution (B) is added to the hot solution. After slightly cooling the solution (for about 3 minutes), 5 drops of solution (C) are added, **without stirring**. A cherry red gold sol results. Two additional drops of solution (B) intensify the color.

Explanation It is important that during the reaction the pH value is kept in the range of 4.3 to 4.5. Depending on the alkaline range of the solution, the gold sol results more or less rapidly.

Red gold sols are produced in an acidic solution, while blue gold sol is formed in an alkaline solution.

Waste Disposal The gold residues are collected separately for recycling.

Spectacular Chemical Experiments. Herbert W. Roesky
Copyright © 2007 WILEY-VCH Verlag GmbH & Co. KGaA, Weinheim
ISBN: 978-3-527-31865-0

Experiment 43 Cherry Red Gold Sol

Experiment 44/43: Left: blue gold sol. Right: red gold sol.

Experiment 44
The Blue Gold

Beauty and affection soon loose their worth,
The value of gold, however, remains.

Johann Wolfgang von Goethe

Apparatus	A 500-mL beaker, one glass rod, 500-mL measuring cylinder, Pasteur pipette, 2-mL pipette, safety glasses, protective gloves.
Chemicals	$HAuCl_4 \cdot xH_2O$ (yellow, about 50% Au content), hydrazine hydrate, distilled water.
Attention!	Safety glasses and protective gloves must be used at all times.
Experimental Procedure	For this demonstration the following solutions are prepared:

Solution (A): 1 g of $HAuCl_4 \cdot xH_2O$ is dissolved in 500 mL of distilled water.
Solution (B): A solution of 1 mL hydrazine hydrate is added to 1000 mL of distilled water. Solution (B) must always be freshly prepared.

2 mL of the gold solution (A) are added to 500 mL of water, and the diluted hydrazine solution (B) is added dropwise at room temperature. A clear blue gold sol results.
  Using a higher concentration of the gold solution yields a darker solution, although this seems to be cloudy.

Explanation Hydrazine hydrate is a very strong reducing agent, and reacts with Au(I)ions to metallic gold according to the following equation:

$$N_2H_4 + 4\,Au^+ \xrightarrow{H_2O} 4\,Au + N_2 + 4\,H_{aq}^+$$

The color of the sol depends on the size of the particles and the pH value of the solution.

Waste Disposal The gold residues are collected separately for recycling.

Spectacular Chemical Experiments. Herbert W. Roesky
Copyright © 2007 WILEY-VCH Verlag GmbH & Co. KGaA, Weinheim
ISBN: 978-3-527-31865-0

Experiment 45
Silver Sol by Electric Discharge

The noble simplicity in nature's work is often due to the noble short-sightedness of the observer.

Georg Christoph Lichtenberg

Apparatus	A Petri dish (11 cm diameter, 6.5 cm high), two silver wires (about 1 mm thick and 20 cm long), two insulation tubes, two alligator clips, two cables, 110 V direct current supply, two stands, bosses, clamps, safety glasses, protective gloves.
Chemicals	Distilled water.
Attention!	Contact with the apparatus after switching on the current should be avoided. A current of 110 V may kill you! Safety glasses and protective gloves must be used at all times.
Experimental Procedure	The two silver wires are placed into tight insulation tubes and connected with the alligator clips to the direct current supply. The Petri dish is then filled with distilled water to a depth of about 4 cm. The silver electrodes are then introduced so that their tops almost touch each other. When the current is switched on, a flash of lightning is produced. Starting out from the reaction source, brown clouds of metallic silver begin to spread all over the water. This experiment can be repeated very often if, after every lightning flash, the top of the electrodes are readjusted.
Explanation	Due to the direct current of 110 V, a voltaic arc develops. Under these temperature conditions, metallic silver vaporizes and dissolves in a colloidal form.
Waste Disposal	The silver residues are collected separately for recycling.

Spectacular Chemical Experiments. Herbert W. Roesky
Copyright © 2007 WILEY-VCH Verlag GmbH & Co. KGaA, Weinheim
ISBN: 978-3-527-31865-0

Experiment 46
How to Make a Silver Sol

Edgar F. Smith, an American student, studied in Göttingen from 1874 to 1876. Smith completed his PhD with Wöhler, and later was for many years president of the University of Pennsylvania in Philadelphia, and president of the American Chemical Society. He used to tell that already during his time in Wöhler's institute the chemical residues were recycled:

"I was working in the laboratory for some weeks when I met Professor Wöhler for the first time. During my practical work I had to produce several pounds of phosphoric chloride. For this purpose I had to generate a considerable amount of chloride, and deposit the residues into the appropriate container. Once when I just was emptying my beaker into the garbage container, somebody tipped on my shoulder. When I looked up I noticed that it was Geheimrat Wöhler. He asked me what I was doing. After having told him all about the chemical preparation, he asked me about the cost of the chemicals I was using. I was not able to calculate them immediately. He then asked me to find this out, and look for a merchant selling chemicals who could use the residues containing the manganese chloride. Soon I discovered that I could sell those provided that they were chemically pure. Wöhler told me to collect the residues separately, and to develop a method for getting a chemically pure product. After thinking for a long time I discovered a method how to separate iron as an impurity, and in this way I could earn enough money to buy additional starting materials for my experiments."

G. Beer, *Göttinger Jahresblätter 1982*

Apparatus — One 400-mL beaker, one 250-mL beaker, glass rods, two 1-mL pipettes, one 5-mL measuring cylinder, 100-mL measuring cylinder, tripod, wire gauze, Bunsen burner, safety glasses, protective gloves.

Chemicals — $AgNO_3$, D(+)glucose, NaOH, distilled water.

Attention! — Avoid skin contact with the silver salt, as it causes black spots on the skin which will take over a week to disappear. Safety glasses and protective gloves must be used at all times.

Spectacular Chemical Experiments. Herbert W. Roesky
Copyright © 2007 WILEY-VCH Verlag GmbH & Co. KGaA, Weinheim
ISBN: 978-3-527-31865-0

Experimental Procedure

0.5 mL of 0.1 M $AgNO_3$ solution is added to 100 mL distilled water in the 400-mL beaker. In the second beaker, 100 mL of distilled water are added to 5 mL D(+)glucose (0.5%) and 1 mL of 2 M sodium hydroxide. The solution contained in the second beaker is then poured into the first beaker. The resulting cloudy solution is slowly warmed, and yields a dark, cloudy silver sol.

Explanation

As early as 1856, J. Liebig used milk sugar for the reduction of silver oxide.

Waste Disposal

The silver residues are always collected separately. They are easily recycled by dissolving the silver residues in concentrated nitric acid and treating the resulting solution with hydrochloric acid (1:1). The silver chloride precipitates, and the left-over solution is decanted to yield AgCl by filtration. The precipitate is placed in a beaker and hydrochloric acid added (1:1). Zinc in the form of a bar is added in order to reduce the Ag^+ ions to metallic silver. The resulting silver precipitate is filtered and washed with distilled water to eliminate chloride and zinc. Finally, concentrated nitric acid is used to convert the metallic silver to $AgNO_3$.

Silver and silver compounds are lethal to microorganisms, and so silver residues should never be flushed down the drain. When silver ions come into contact with the skin, the skin becomes brown or black due to the reduction of metallic silver. Silver nitrate is used to remove warts. In this case, the $AgNO_3$ is reduced, thereby forming silver and caustic nitric acid.

Reference

J. Liebig, *Liebigs Ann. Chem.* **1856**, *90*, 132.

Experiment 47
The Reaction of Silver Nitrate with Tannin

The reward of the scientist is what Henri Poincaré calls the pleasure of understanding, and not the production for a particular use.

Albert Einstein

Apparatus	A 250-mL beaker, glass rod, 2-mL pipette, two dropping pipettes, tripod, wire gauze, Bunsen burner, safety glasses, protective gloves.
Chemicals	0.1 M $AgNO_3$ solution, soda solution (1%), tannin solution (1%), distilled water.
Attention!	Avoid skin contact with the silver salt as it causes black spots on the skin. Safety glasses and protective gloves must be used at all times.
Experimental Procedure	2 mL of the 0.1 M silver nitrate solution are poured into the beaker, and distilled water is added to give a total volume of 100 mL. Five drops of a freshly prepared tannin solution (1%) and one drop of a soda solution (1%) are added. The solution turns yellow. Warming the solution yields a yellow to red brown colored sol.
Explanation	The color of the silver sol depends on the particle size. The smaller the particles, the stronger is the coloring power.
Waste Disposal	The silver residues are collected separately for recycling.

Part V
Fascinating Experiments by Self-Organization

Those natures which, when they meet, quickly lay hold on and mutually affect one another we call affined. This affinity is sufficiently striking in the case of alkalis and acids which, although they are mutually antithetical, and perhaps because they are so, most decidedly seek and embrace one another, modify one another, and together form a new substance. Think only of lime, which evidences a great inclination, a decided desire for union with acids of every kind. As soon as our cabinet of chemicals arrives we will show you some very entertaining experiments which will give you a better idea of all this than words, names and technical terms.

". . . the affinities become interesting only when they bring about divorces." "Does that doleful word, which one unhappily hears so often in society these days, also occur in natural science?" Charlotte exclaimed. "To be sure," Eduard replied, "it even used to be a title of honor to chemists to call them artists in divorcing one thing from another." "Then it is not so any longer," Charlotte said, "and a very good thing too. Uniting is a greater art and a greater merit. An artist in unification in any subject would be welcomed the world over."

". . . I know, alas, of all too many cases in which an intimate and apparently indissoluble union between two beings has been broken up by a chance association with a third and one of the couple at first so fairly united driven out into the unknown." "Chemists are far more gallant in this matter," said Eduard: "they introduce a fourth, so that no one shall go empty away." "Yes, indeed!" the Captain added: "these cases are in fact the most significant and noteworthy of all; in them one can actually demonstrate attraction and relatedness, this as it were crosswise parting and uniting: where four entities, previously joined together in two pairs, are brought into contact, abandon their previous union, and join together afresh. In this relinquishment and seizing, in this fleeing and seeking, one really can believe one is witnessing a higher determination;. . . ."

Johann Wolfgang von Goethe, *Elected Affinities* (Penguin Classics)

Spectacular Chemical Experiments. Herbert W. Roesky
Copyright © 2007 WILEY-VCH Verlag GmbH & Co. KGaA, Weinheim
ISBN: 978-3-527-31865-0

Experiment 48
Dissipative Structures: Chemical Patterns in Aqueous Solution

"And of course the spoken word only creates a shadow of a picture, only shows the student the outline of a science, which becomes understandable and colorful only when it is presented by an experiment."

Karl Heumann, *Anleitung zum Experimentieren bei Vorlesungen über anorganische Chemie zum Gebrauch an Universitäten und Technischen Hochschulen sowie beim Unterricht an höheren Lehranstalten.* (1876)

Apparatus	One glass dish (11.5 cm diameter, 1.5 cm high), filter paper (15 cm diameter), 50-mL measuring cylinder, Pasteur pipette, plastic bag with a clip, overhead projector, safety glasses, protective gloves.
Chemicals	Iodine dissolved in ethyl alcohol (10%), 0.1 M KI solution, starch solution (0.5%).
Attention!	Iodine vapors are dangerous for eyes and mucous membranes. Safety glasses and protective gloves must be used at all times.
Experimental Procedure	A filter paper is sucked in the center with about 0.5 mL of the iodine solution. After a short time the ethyl alcohol evaporates and the iodine remains on the filter paper.

This procedure is carried out a second time. The thus prepared filter paper can be kept in a closed plastic bag for several hours until the experiment is carried out.

The glass dish is placed on the overhead projector and filled with 10 mL of the starch solution (0.5%) and 40 mL of the KI solution. Now, the prepared filter paper is placed on the brim of the glass dish. After about 1 minute the filter paper is removed. A form of cloudy pattern can be seen which slowly turns from brown to deep blue in color. With an empty Pasteur pipette small air turbulences can be produced that change these patterns.

Spectacular Chemical Experiments. Herbert W. Roesky
Copyright © 2007 WILEY-VCH Verlag GmbH & Co. KGaA, Weinheim
ISBN: 978-3-527-31865-0

Explanation

With the starch solution, free iodine forms the blue-black amylose iodide complex containing I_5^-. The amylopectin that is also present in the starch solution reacts with iodine to give a purple solution. In concentrated aqueous alkali, the iodide-iodine solution turns brown. In the case of a strongly diluted solution, the brown color turns into yellow due to the I_3^- ion.

Formation of the dissipative structures continues until the chemicals are completely consumed, and equilibrium is reached. The structures are formed on the phase boundary between gas and liquid. The resulting products are not solids precipitated during the reaction but are soluble in the aqueous phase, so that from diffusion or stirring a homogeneous solution results.

The patterns show different structures and shapes. Thus, they are attractive and stimulating to realize various experiments.

Waste Disposal

The solution in the glass dish can be flushed down the drain.

References

- D. Avnir, M.L. Kagan, *Chaos* **1995**, *5*, 589.
- G. Nicolis, J. Prigogine, *Self-Organization in Non-Equilibrium Systems*, Wiley, New York, 1977.
- D. Avnir, M. Kagan, *Naturwissenschaften* **1983**, *70*, 361.
- R.C. Teitelbaum, S.L. Ruby, T.J. Marks, *J. Am. Chem. Soc.* **1978**, *100*, 3215.

Experiment 48 Dissipative Structures: Chemical Patterns in Aqueous Solution | 117

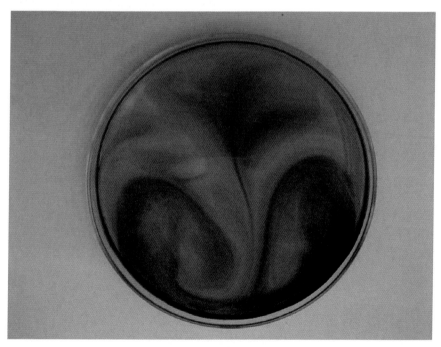

Experiment 48: The iodine-starch reaction. On top: I_5^- ions. At the bottom: I_3^- and I_5^- ions.

Experiment 49
Acidic Acid Butyl Ester in the Presence of Bromocresol Green

A great genius will seldom make his discoveries on paths frequented by others. When he discovers things he usually also discovers the path to discovery.

Georg Christoph Lichtenberg

Apparatus One Petri dish (11 cm diameter, 1.5 cm high), one Petri dish of 16 cm diameter, filter paper (15 cm diameter), paper towels, overhead projector, one 250-mL beaker, 1-mL pipette, 100-mL measuring cylinder, safety glasses, protective gloves.

Chemicals Bromocresol green (sodium salt of the 3', 3", 5', 5"-tetrabromo-m-cresol-sulfonophthalein), sodium hydroxide, acidic acid butylester.

Attention! Safety glasses and protective gloves must be used at all times.

Experimental Procedure Before running this experiment, 1 mL of bromocresol green solution (1%) is added to 100 mL 0.005 M sodium hydroxide solution, which leads finally to a blue solution. 50 mL of this solution are poured into the first Petri dish, which is placed on the overhead projector. In the second Petri dish, the filter paper is wetted with acidic acid butylester, and any excess ester on the filter paper is removed by dabbing with the paper towel. The filter paper is placed for 2 minutes on the brim of the Petri dish containing the bromocresol green solution. After removing the filter paper, yellow spots appear in the blue solution, which slowly turns green.

Explanation The acidic acid butyl ester reacts in aqueous solution to give acidic acid and butyl alcohol. The presence of the acidic acid is shown by the color change of bromocresol green from blue to yellow. In contrast to acidic acid butyl ester, under these conditions no ester cleavage can be demonstrated with acidic acid ethyl ester.

Waste Disposal The residues can be flushed down the drain.

Spectacular Chemical Experiments. Herbert W. Roesky
Copyright © 2007 WILEY-VCH Verlag GmbH & Co. KGaA, Weinheim
ISBN: 978-3-527-31865-0

Experiment 50
Precipitation Using the Gas Phase

In a report published in the *Göttinger Taschenkalender* of 1783, Lichtenberg's experiments in his main lecture were described:

"We are dealing (1) with the production of atmospheric air, specially in its pure state, because it can be called dephlogisticated air (O_2), (2) the so called fixed air (CO_2), (3) the combustible air (H_2), (4) the nitric air (a mixture of NO and NO_2), (5) the vitriol acidic air, better sulfuric air (SO_2), (6) the hydrochloric acidic air (HCl), (7) the acidic acid air, (8) the ammonia air (NH_3) and (9) the fluorspar acidic air (H_2F_2 polluted by H_2SiF_6).
With these different gases or, as Lichtenberg said, kinds of air, many impressing experiments could be realized. Lichtenberg was far too much a researcher than not to be fascinated by these possibilities. His pleasure was much more than scientific interest.
In his club he banged with explosive gas, he wanted his friends to deal with this delightful object, and he wrote them long letters with detailed instructions, he sent them dephlogisticated air in bottles. He did this so often that the bottles went to and fro between Göttingen and Hanover like buckets in a well."

Apparatus Glass dish (9 cm diameter, 2 cm high), 50-mL measuring cylinder, filter paper (12.5 cm diameter), Pasteur pipette, overhead projector, safety glasses, protective gloves.

Chemicals 0.01 M $CuSO_4$ solution, ammonia solution (25%, about 13 M).

Attention! Ammonia solutions are irritating to skin and eyes, and the vapor must not be inhaled. If ammonia is splashed on the skin, it must be washed off with plenty of water! Safety glasses and protective gloves must be used at all times.

Experimental Procedure 50 mL of a 0.01 M $CuSO_4$ solution are poured into the glass dish, which is placed on the overhead projector. The solution appears colorless, both on the projector and projected onto the screen. A filter paper is then sucked with concentrated ammonia solution by means of the pipette. The ammonia solution is applied circularly starting from the center of the filter paper. When the impregnated part of the filter paper has a diameter of 5–7 cm, it is placed on the brim of the dish for about 1 minute. When the filter paper has been removed, a precipitate of basic copper sulfate appears. The picture shows dissipative, well-formed patterns.

On the screen, light brown stripes can be seen, while the solution in the background seems to be colorless.

However, the solution in the glass dish placed on the projector is light blue and the precipitate is white.

Explanation

Ammonia reacts with Cu(II) ions to give $Cu(OH)_2$, which dissolves in excess ammonia to form the dark blue $[Cu(NH_3)_4]^{2+}$ complex.

Waste Disposal

The copper residues are collected and placed in the container used for heavy metals.

Experiment 51
Methods Become Accepted: Nessler's Reagent and Gaseous Ammonia

"And Socrates sat down and said: It would be nice, Agathon, if, when we touch each other, wisdom would flow from the fuller into the more empty one, like water in beakers which flows through a thread from the fuller beaker into the more empty one."

Platon

Apparatus   Glass dish (9 cm diameter, 2 cm high), 50-mL measuring cylinder, filter paper (12.5 cm diameter), Pasteur pipette, overhead projector, safety glasses, protective gloves.

Chemicals   Nessler's reagent ($HgCl_2$ solution, 2%), KI solution (10%), potassium hydroxide solution (50%), ammonia solution (medium concentrated, about 7 M), distilled water.

Attention!   Mercury and mercury compounds are extremely toxic. Safety glasses and protective gloves must be used at all times.

Experimental Procedure   Nessler's reagent is freshly prepared immediately before the demonstration. To 5 mL of a $HgCl_2$ solution (12%), 10 mL of a KI solution (10%) are added. First, a red precipitate is formed which afterwards dissolves. Then, 15 mL of a potassium hydroxide solution (50%) are added, and distilled water is added to a final volume of 100 mL. 50 mL of this solution are now poured into the glass dish placed on the overhead projector. The filter paper is dipped into a medium concentrated ammonia solution (about 7 M). The impregnated filter paper is then placed on the brim of the glass dish for about 15 seconds. After having removed the filter paper, beautifully formed patterns appear which are grey black on the screen and yellow brown in solution.

Medium concentrated ammonia solution has been used because this solution does not have as pungent and unpleasant a smell as the concentrated ammonia solution.

Spectacular Chemical Experiments. Herbert W. Roesky
Copyright © 2007 WILEY-VCH Verlag GmbH & Co. KGaA, Weinheim
ISBN: 978-3-527-31865-0

Explanation Mercury(II) dichloride reacts with KI to yield the complex $K_2[HgI_4]$. A solution of this complex mixed with KOH is known as "Nessler's reagent", and serves for the sensitive identification of ammonia. The yellow-brown precipitate is $[Hg_2N]I$, and is a derivative of Millon's base $[Hg_2N]OH$.

Waste Disposal The mercury-containing solutions are collected in the containers used for heavy metals.

Experiment 52
Reduction of KMnO₄ with Ethyl Alcohol

The target of natural sciences is on the one hand to understand, if possible, completely the relationship between perceptible experiences as a whole and on the other hand to reach this target using a minimum of basic principles and basic relationships.

Albert Einstein

Apparatus One Petri dish (9–10 cm diameter, 1.5 cm high), one dish (15 cm diameter), one filter paper (12.5 cm diameter), one 100-mL beaker, one 50-mL beaker, 2-mL measuring cylinder, glass rod, Pasteur pipette, overhead projector, paper towels, safety glasses, protective gloves.

Chemicals 1 M $KMnO_4$, 18 M H_2SO_4, ethyl alcohol, distilled water.

Attention! Potassium permanganate solution is a powerful oxidant. Skin contact should be avoided. Sulfuric acid is highly dangerous; it causes severe burns to skin and eyes. Safety glasses and protective gloves must be used at all times.

Experimental Procedure 10 drops of the 1 M $KMnO_4$ solution are added to 50 mL of distilled water. 20 mL of this solution are mixed with 2 mL of 18 M sulfuric acid. The resultant purple solution is poured into the Petri dish which is placed on the overhead projector. A filter paper is then dipped into ethyl alcohol, dried slightly between paper towels, and placed on the brim of the glass dish. After 2 minutes the filter paper is removed. The solution becomes colorless, which can be seen by large colorless spots in the reddish purple solution.

Explanation In an acidic solution the ethyl alcohol reduces the permanganate to produce manganese(II). During this reduction, the color changes from reddish purple to almost colorless, because it is impossible to

Spectacular Chemical Experiments. Herbert W. Roesky
Copyright © 2007 WILEY-VCH Verlag GmbH & Co. KGaA, Weinheim
ISBN: 978-3-527-31865-0

recognize the color of the manganese(II) at such a low concentration.

Waste Disposal The residue of the solution in the Petri dish is neutralized with sodium hydroxide, and the resulting solution can be flushed down the drain.

Experiment 52 Reduction of KMnO4 with Ethyl Alcohol

Experiment 52: Reduction of a KMnO$_4$ solution with ethyl alcohol. On top: KMnO$_4$ solution. At the bottom: partly reduced KMnO$_4$ solution.

Experiment 53
Alcohol Test

Like a chemist Napoleon regards Europe as to be the material for his experiments. But very soon this material reacted against him.

Frederic Bastiat, *The Law*

Apparatus One glass dish (9 cm diameter, 1.5 cm high), one filter paper (12.5 cm diameter), one 25-mL beaker, one 10-mL and one 15-mL measuring cylinders, glass rod, one dish (glass or plastic, about 15 cm diameter), paper towels, overhead projector, safety glasses, protective gloves.

Chemicals $K_2Cr_2O_7$, 18 M H_2SO_4, C_2H_5OH (97%).

Attention! $K_2Cr_2O_7$ is highly toxic. In general, chromium in high oxidation states is mutagenic.

Sulfuric acid causes severe burns to skin and eyes; hence, the experiment should be carried out in a well-ventilated fume hood. If sulfuric acid comes into contact with the skin, it should be washed off quickly with plenty of water. Safety glasses and protective gloves must be used at all times.

Experimental Procedure 5 mL of concentrated H_2SO_4 are added under stirring to 10 mL of the $K_2Cr_2O_7$ solution (5%). The hot solution is poured into the glass dish placed on the overhead projector. In the bigger glass dish, a filter paper is sucked with ethyl alcohol, dried slightly between paper towels, and placed on the brim of the glass dish containing the chromium solution. After 1 minute the filter paper is removed, whereupon the green color of the resulting chromium(III) appears clearly in the orange yellow solution, which after some time turns yellow-green.

The demonstrator is able to see the emerging green color through the filter paper.

A dichromate solution of 2% can also be used, but the color changes are better seen when a solution of 5% is used.

Spectacular Chemical Experiments. Herbert W. Roesky
Copyright © 2007 WILEY-VCH Verlag GmbH & Co. KGaA, Weinheim
ISBN: 978-3-527-31865-0

Explanation

It is very important to prepare the solution of $K_2Cr_2O_7$ and the sulfuric acid immediately before the demonstration. The resulting hot solution oxidizes C_2H_5OH much more easily to give acetaldehyde (CH_3CHO). There, the Cr(VI) is reduced to Cr(III).

Waste Disposal

The chromium-containing solutions are collected in a special container used for highly toxic metal residues.

Experiment 53: On top: $K_2Cr_2O_7$ solution. At the bottom: partly reduced $K_2Cr_2O_7$ solution.

Experiment 54
An Old Hat with New Feathers: the Precipitation of AgCl with HCl Gas

Innumerable germs of the spiritual life fill the space. But only in some rare spirits they find the soil to grow. Inside those the idea of which nobody knows from where it originates will come to life through creativity.

Justus von Liebig

Tomorrow and after tomorrow I will like other paintings, I will paint other paintings.

Hermann Hesse, *Klingsor's last summer*

Apparatus	Glass dish (9 cm diameter, 2 cm high), 50-mL measuring cylinder, filter paper (12.5 cm diameter), Pasteur pipette, overhead projector, paper towels, safety glasses, protective gloves.
Chemicals	0.001 M $AgNO_3$ solution, concentrated hydrochloric acid (12 M, 37%).
Attention!	Contact of the skin with silver salts should be avoided. Black spots may appear on the skin, and these may take over a week to disappear. Safety glasses and protective gloves must be used at all times.
Experimental Procedure	50 mL of 0.001 M $AgNO_3$ solution are poured into the glass dish, which is placed on the overhead projector. A filter paper is soaked with concentrated hydrochloric acid using a Pasteur pipette, dried slightly between paper towels, and placed on the brim of the dish for about 5–10 seconds. After having removed the filter paper, an AgCl precipitate appears. A well-structured pattern is formed, instead of a white precipitate, as occurs when using a test tube for the experiment.
Explanation	The AgCl precipitate is formed through the hydrogen chloride supplied through gas phase and the silver ions present in the solution.

Spectacular Chemical Experiments. Herbert W. Roesky
Copyright © 2007 WILEY-VCH Verlag GmbH & Co. KGaA, Weinheim
ISBN: 978-3-527-31865-0

134 *Experiment 54 An Old Hat with New Feathers: the Precipitation of AgCl with HCl Gas*

Waste Disposal   The silver-containing solutions are collected separately in the container used for silver residues. Silver-containing solutions should under no conditions be flushed down the drain.

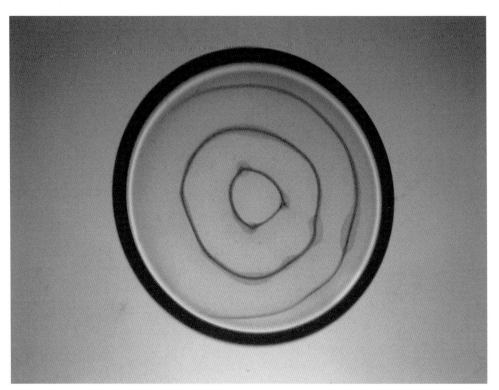

Experiment 54: Precipitation of AgCl with HCl over the gas phase.

Part VI
Chemical Varieties

For the inventor Schönbein, the gun cotton was first a simple joke article. He just had fun to demonstrate the bang to his friends. He often was invited to attend some evening invitation, because he was a fantastic entertainer. He always carried a small box with him where he kept some prepared gun cotton. To amuse the guests after dinner, he ignited it with a candle producing a big bang. His students in the lecture hall also were fascinated and laughing and even people at the Prince's court were amused. Soon he was imitated at the fair. The market people did not want to miss this attraction. Even some municipal theatres performed a comedy called "The exploding cotton". The principal actress wore a dress made of gun cotton, which exploded at the end of the performance. The bang did not cause any damage, but after the explosion the actress was dressed in Eve's suit. Thunderous applause.

E.F. Schwenk, *Sternstunden der frühen Chemie* (*Great Moments of the Early Days of Chemistry*), H.C. Beck, München, 1998.

Experiment 55
A Chemical Buoy

If there is a way to make things better, find it.

Thomas Alva Edison

Apparatus A 2-L flat-bottomed flask, magnetic stirrer, funnel, two 400-mL beakers, one 800-mL beaker, 50-mL measuring cylinder, three 1-L measuring flasks, safety glasses, protective gloves.

Chemicals H_2O_2 (35%), KSCN, $CuSO_4 \cdot 5H_2O$, NaOH, luminol, distilled water.

Attention! H_2O_2 reacts explosively with organic materials. It can cause burns; consequently, skin contact must be avoided. Safety glasses and protective gloves must be used at all times.

Experimental Procedure The following solutions are prepared in the measuring cylinders:

Solution (A): 50 mL of H_2O_2 (35%).
Solution (B): (0.15 M KSCN) 14.55 g KSCN are dissolved with water to a total volume of 1 L.
Solution (C): (6×10^{-4} M $CuSO_4$) 0.15 g $CuSO_4 \cdot 5H_2O$ are dissolved in water to a total volume of 1 L.
Solution (D): (0.10 M NaOH and 3.7×10^{-3} M luminol) 4 g of solid NaOH are dissolved in 100 mL of water, and 0.55 g of luminol is added while stirring. The solution is then diluted with water to a total volume of 1 L.

The flat-bottomed flask is placed on the magnetic stirrer, and 300 mL of solution (B), 600 mL of solution (C), 300 mL of solution (D), and 50 mL of solution (A) are poured into the flask. The magnetic stirrer is switched on in order to obtain a homogeneous solution. The room is then completely darkened. After about 1 minute the mixture begins to shine very weakly for about 2–3 seconds, but after that it can be seen very distinctly. This reaction continues for about 3 to 10 minutes,

Spectacular Chemical Experiments. Herbert W. Roesky
Copyright © 2007 WILEY-VCH Verlag GmbH & Co. KGaA, Weinheim
ISBN: 978-3-527-31865-0

with a shining interval of 60 seconds which slowly increases. The experiment is completed after about 20 minutes.

Explanation

Oscillating chemical reactions (e.g., the Belousov–Zhabotinsky reaction) are very impressive experiments for demonstrations, as the spectators are fascinated by the spontaneous color changes.

By adding luminol, a slightly light-emitting reaction takes place which is still more fascinating for the spectators. The oscillating frequency depends essentially on the concentration and temperature of the system.

The experimental procedure showed that the luminescence would be much more perceptible, if a concentrated H_2O_2 solution were to be used instead of a diluted one.

Waste Disposal

When the experiment is finished, the solutions are concentrated on a water bath. The residues are collected in the container used for heavy metals.

References

- H.E. Prypsztejn, *J. Chem. Educ.* **2005**, *82*, 53–54.
- F.W. Schneider, J. Amrhen, *J. Phys. Chem.* **1988**, *92*, 3318–3320.

Experiment 56
Flower Power

We are now starting to talk about philosophers. I hope you will remember that every time you got a result, especially a new one, you should ask yourself: "Why is it like that? Why does it happen?". Then gradually you will find the reason.

Michael Faraday

Apparatus	Petri dish (10 cm diameter), 10-mL and 25-mL measuring cylinders, knife, a pair of tweezers, overhead projector, safety glasses, protective gloves.
Chemicals	Bromothymol blue, phenolphthalein, *n*-propanol and *iso*-propanol, sodium.
Attention!	Sodium reacts explosively with water. Safety glasses and protective gloves must be used at all times.
Experimental Procedure	First, a solution of 5 mg bromothymol blue, 20 mg phenolphthalein, 25 mL *n*-propanol and 25 mL *iso*-propanol is prepared. About 10 mL of this solution are poured into the Petri dish, and a piece of sodium weighing about 100 mg is placed in the center of the dish. For this amount of liquid the buoyancy for the sodium piece is too small, so that it stays in the center of the dish. A blue-purple flower is then seen to grow from below the piece of sodium.
Explanation	The color change is due to the pH value of the solution. In the alkaline region, bromothymol blue turns deep blue. The rate of pattern formation depends on the ratio of the amount of the two alcohols. The reaction of *n*-propanol with sodium is much faster compared to that of *iso*-propanol.
Waste Disposal	The solution in the Petri dish is neutralized with hydrochloric acid and then flushed down the drain.
Reference	W. Ruf, R. Full, *PdN-Chemie* 2000, *49*, 19.

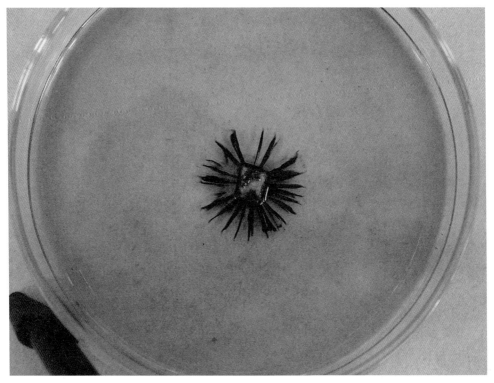

Experiment 56: The reaction of sodium in the presence of *n*- and *iso*-propanol and bromothymol blue.

Experiment 57
Münchhausen: The Flying Styrofoam Ball

"Not to tackle the draft before being content with the whole system encourages and makes work easier."

Georg Christoph Lichtenberg

When Wilhelm Ostwald saw a simple balance in the workroom of Berzelius in Stockholm, he thought: "I understood and I will never forget that it is not the instrument which is important, but the man who uses it. This experience strengthened the tendency, taken along from childhood, to settle with the simplest tools."

Apparatus Plexiglas tube (about 50 cm long, inside diameter 10 cm, wall ca. 0.5 cm thickness; closed on one side with a lid 2.5 cm thick), metal stand with a clamp, Styrofoam ball (ca. 10 cm diameter, with a center hole of ca. 0.8 cm), metal tube (ca. 60 cm long, outer diameter 8 mm), small ventilator (diameter <4 cm, operated with a DC current), a spark inductor with two cables, a rope with a jacketed wire (50 m long), clamp, syringe (50 mL), small PVC vessel, safety glasses, and protective gloves

Chemicals Pentane (boiling point 36 °C).

Attention! The rope must be fixed far above the heads of the spectators. Mixtures of air and pentane are explosive! Safety glasses and protective gloves must be used at all times.

Experimental Procedure The 60-cm metal tube is placed in the center of the Plexiglas tube. A rope of about 50 m length is pulled through the metal tube to control the direction of the flying Styrofoam ball. Inside the lid is: (a) the small ventilator (operated with a 4 V DC motor); (b) two electrodes connected to the spark inductor; and (c) a screw lock (ca. 1 cm) for the injection of pentane.

The Plexiglas tube is attached to the stand with the clamp. The Styrofoam ball is placed in the center of the Plexiglas tube, about

Spectacular Chemical Experiments. Herbert W. Roesky
Copyright © 2007 WILEY-VCH Verlag GmbH & Co. KGaA, Weinheim
ISBN: 978-3-527-31865-0

Experiment 57 Münchhausen: The Flying Styrofoam Ball

20 cm away from the open end of the tube. One end of the rope is fixed carefully to the stand, while the other end is attached to the wall of the room at a distance of about 30–50 m. After having plugged in the spark inductor and the ventilator respectively, about 0.2–0.3 mL of pentane are injected through the lid of the Plexiglas tube into the PVC vessel, and the screw lock is closed. After some minutes the ventilator is switched on in order to mix the pentane vapor with the air. Then, the spark inductor is switched on to produce sparks. The pentane–air mixture ignites due to the sparks, and the resulting explosion drives the ball out of the tube, causing it to run extremely fast along the rope. The set-up of the experiment is shown in the figure.

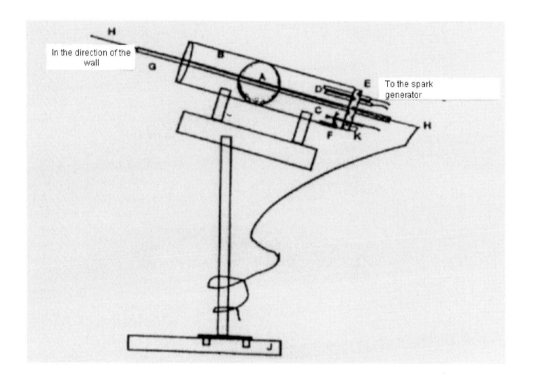

A: Styrofoam ball
B: Plexiglas tube
C: Ventilator
D: Electrode
E: Plastic lid

F: PVC vessel
G: Metal tube
H: Rope with jacketed wire
J: Stand with clamp
K: Screw lock

Experiment 57: Experimental set-up of the flying Styrofoam ball.

Explanation

This arrangement is very safe, because controlled tracking of the ball avoids any collisions with the spectators.

The pentane vapor must be well-mixed with air in order to be in the explosive range of 1.4% and 8.3% pentane. Beyond this range no real reaction takes place. The mixture within the explosion range is ignited by the spark, and during the combustion simple molecules such as CO_2 and H_2O are formed. This reaction is exothermic, and the resulting explosion drives the Styrofoam ball along the rope through the room. Under complete combustion, the reaction occurs according to the following equation:

$$C_5H_{12} + 8\ O_2 \text{ (from the air)} \rightarrow 5\ CO_2 + 6\ H_2O$$

Experiment 58
The Remarkable Rocket

If you are going to build something in the air it is always better to build castles than houses of cards.

Georg Christoph Lichtenberg

"How fortunate it is for the King's son," the Rocket remarked, "that he is to be married on the very day on which I am to be let off. Really, if it had been arranged beforehand, it could not have turned out better for him; but Princes are always lucky."
"Dear me!" said the little Squib, "I thought it was quite the other way, and that we were to be let off in the Prince's honour."
"It may be so with you," he answered; "indeed, I have no doubt that it is, but with me it is different. I am a very remarkable Rocket, and come of remarkable parents. My mother was the most celebrated Catherine Wheel of her day, and was renowned for her graceful dancing. When she made her great public appearance she spun round nineteen times before she went out, and each time that she did so she threw into the air seven pink stars. She was three feet and a half in diameter, and made of the very best gun powder. My father was a Rocket like myself, and of French extraction. He flew so high that the people were afraid that he would never come down again. He did, though, for he was of a kindly disposition, and he made a most brilliant descent in a shower of golden rain. The newspapers wrote about his performance in very flattering terms. Indeed, the Court Gazette called him a triumph of Pylotechnic art."
"Pyrotechnic, Pyrotechnic, you mean," said a Bengal Light; "I know it is Pyrotechnic, for I saw it written on my own canister."

. . . . . .

"I am made for public life," said the Rocket, "and so are all my relations, even the humblest of them. Whenever we appear we excite great attention. I have not actually appeared myself, but when I do so it will be a magnificent sight. As for domesticity, it ages one rapidly, and distracts one's mind from higher things." . . .
The Rocket was very damp, so he took a long time to burn. At last, however, the fire caught him.
"Now I am going off!" he cried, and he made himself very stiff and straight. "I know I shall go much higher than the stars, much higher than the moon, much higher than the sun. In fact, I shall go so high that——"
Fizz! Fizz! Fizz! And he went straight up into the air.
"Delightful!" he cried, "I shall go on like this for ever. What a success I am!"

Spectacular Chemical Experiments. Herbert W. Roesky
Copyright © 2007 WILEY-VCH Verlag GmbH & Co. KGaA, Weinheim
ISBN: 978-3-527-31865-0

Experiment 58 The Remarkable Rocket

But nobody saw him.
Then he began to feel a curious tingling sensation all over him.
"Now I am going to explode," he cried. "I shall set the whole world on fire, and make such a noise that nobody will talk about anything else for a whole year."
And he certainly did explode. Bang! Bang! Bang! Went the gunpowder. There was no doubt about it.
But nobody heard him, not even the two little boys, for they were sound asleep. Then all that was left of him was the stick, and this fell down on the back of a Goose who was taking a walk by the side of the ditch.
"Good heavens!" cried the Goose. "It is going to rain sticks"; and she rushed into the water.
"I knew I should create a great sensation," gasped the Rocket, and he went out.

Oscar Wilde, *Complete Fairy Tales*

Apparatus

A 25- to 30- meter length of nylon rope (a plastic clothesline is fine), a 1-L polyethylene terephthalate (PET) bottle (e.g., Coca Cola bottle), one 100-mL syringe, one stop watch, a 1-mL syringe with a cannula of 4–5 cm, PVC tube (15 cm long, ca. 1 cm in diameter), one spark inductor with two cables (each 2 m long), one large stand (about 1.20 m high), one screw clamp, one clamp, two copper electrodes (in a support which fits into the bottle neck, and which is provided with a hole for the cannula of the syringe), safety glasses, protective gloves.

Chemicals

Pentane (0.3 mL per experiment; boiling point 36 °C), oxygen (gas, 100 mL per experiment).

Attention!

Pentane, when mixed with air and oxygen, forms explosive mixtures!
The amounts of chemicals used must be EXACTLY those indicated, otherwise an explosion may occur. The experiment should never be run under pure oxygen. Earplugs should be used because of the big bang.
Safety glasses and protective gloves must be used at all times.

Experimental Procedure

The PVC tube is fixed with Scotch or a similar tape in the middle of the long side of the bottle. This construction acts as the tracking for the 1-L PET bottle along the rope. One end of the rope is fixed to the wall of the lecture hall. It must be fixed very high, so that nobody can touch it. The rope is passed through the PVC tube fixed on the bottle, and its other end is fixed on the stand, which is fastened at the table with the screw clamp. The bottle neck is directed to the stand (see illustration). The electrodes are fixed on the stand with a clamp. 100 mL of oxygen is filled into the PET bottle by means of the syringe.

The bottle is fixed on the electrode support. Pentane (0.3 mL) is then injected with the syringe through the additional hole of the electrode support into the PET bottle. After about 2–3 minutes, the liquid pentane has vaporized. The resulting oxygen–air–pentane mixture is now reacted with the spark inductor. The spark produced between the electrodes ignites the gas mixture, yielding a bang. The bottle (which is fixed on the rope) runs along the rope to its other end, producing a blue flame as it is propelled.

This experiment can be repeated after refilling of the PET bottle with pentane/oxygen.

Reference   H.W. Roesky, *PdN-Chemie* **2000**, *49*, 2.

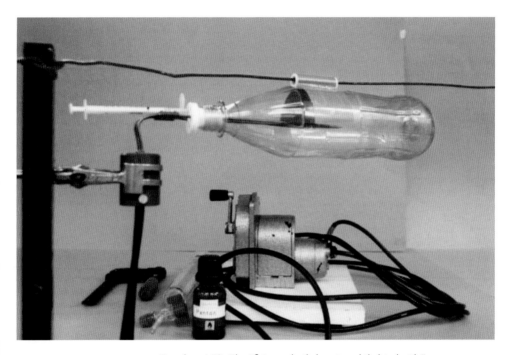

Experiment 58: The "flying polyethylene terephthalate bottle".

Experiment 59
Eatable Burning Banana

This story was told by my brother's wife, and it happened close to Borås. Her girl friend had bought some bananas, she gave one to her little boy, and told him to sit down and eat it. Then she went to the kitchen. And then she heard him call: "Mama, something is moving in the banana!"
"Yes, yes, of course", said the mother, "the banana is alife!"
She thought that he was dreaming. But when she entered the room, he was lying on the floor, and he was dead. He had been bitten by a small snake, four to five centimeters long, a kind of mini cobra. Since I heard this story I always look into the banana before I eat it."

Klintberg 1990

Apparatus	Candle holder, knife, lighter, safety glasses, protective gloves.
Chemicals	Banana, pecan kernel (oily).
Attention!	Safety glasses and protective gloves must be used at all times.
Experimental Procedure	A large banana is peeled, and cut diagonally into two pieces. One of the pieces is placed in the candle holder. A piece of the pecan kernel is put into the banana. The nut is then ignited with the lighter. The courageous demonstrator can now bite a piece off the burning banana.

Spectacular Chemical Experiments. Herbert W. Roesky
Copyright © 2007 WILEY-VCH Verlag GmbH & Co. KGaA, Weinheim
ISBN: 978-3-527-31865-0

Experiment 60
Burning Pecan

One of our friends told us a story which friends of her parents had told her: "After the war the family of these friends got regularly care parcels from their American relatives. Chocolate, coffee, milk powder, all these things had been sent. One day, a parcel with a big black box inside arrived. The family opened it to see what was in there, and they found that it contained a powder which they thought should be – especially because of the precious packing – the newest nutrition supplements. The box was put in the kitchen, and every day the whole family enjoyed the special nutrition supplement. Some weeks later, a letter arrived. "Eight weeks ago our beloved mother and grandmother passed away. In accordance with her instructions we sent her ashes to her German home country. The urn will arrive in the next few days. Please, bury her solemnly. Enclosed we are sending you 200 dollars to cover your expenses."

Brednich (1999)

Apparatus	One large, slightly meltable test tube (ca. 3 cm diameter, 20 cm high), stand, boss, Bunsen burner, safety glasses, protective gloves.
Chemicals	KNO_3, pecan kernel (ca. 1 g).
Attention!	The heating of KNO_3 should be avoided in the presence of materials which can easily be oxidized. Safety glasses and protective gloves must be used at all times.
Experimental Procedure	7 g of KNO_3 are placed in a test tube and heated to a clear melt until foam is produced. The pecan kernel is then added to the salt melt. The kernel reacts under oxidation and flames are generated. However, when the reaction occurs with incomplete combustion, dense smoke is formed. The procedure can be modified by retaining a piece of the nut shell, when a violent reaction takes place, even producing fire on occasion.

Spectacular Chemical Experiments. Herbert W. Roesky
Copyright © 2007 WILEY-VCH Verlag GmbH & Co. KGaA, Weinheim
ISBN: 978-3-527-31865-0

Explanation When heated, KNO_3 decomposes into KNO_2 and oxygen. This is shown by the formation of oxygen bubbles in the clear melt.

Waste Disposal After cooling the test tube to room temperature, the remaining KNO_3 and KNO_2 can be dissolved in water and flushed down the drain.

Experiment 61
Sparks and Shining Fire

See how it gleams! – Now we may hope to see
Results. The ingredients – our manifold
Materia anthropica, they are called –
We mix in a retort most patiently,
With all due care, and so by perlutation
They quietly reach their consummation.

Johann Wolfgang von Goethe, *Faust*

In 1827, Friedrich Wöhler was able to produce pure aluminum by reduction of aluminum trichloride with potassium. On the occasion of his first visit to Berzelius in Stockholm, Wöhler reported:

"When he showed me his laboratory", Wöhler tells, "I thought I was dreaming, I could not believe that I really was in these classical rooms. The laboratory was next to the living room. It consisted of two normal rooms with very simple furniture; there was no stove neither a hood, no water or gas supply. In one room there were two simple tables made out of pine wood; one of them was Berzelius' desk, the other one was mine. Some cupboards containing some few reagents were standing on the walls. There was not a big choice. Because when I needed potassium prussiate for doing my experiments I had to order it in Lübeck. In the center of the room were the mercury trough and the table for glass blowing, the latter one was placed under a chimney hood of wax taffeta leading into the chimney of a normal stove. The sink consisted of a stoneware water recipient with a water tap and a pot under it. In the other room were balances and other instruments, close by a small work shop with a lathe. In the kitchen where Anna, cook and factotum of the master who at that time was not yet married, prepared the dinner, stood a small annealing furnace and the permanently heated sand bath."

Apparatus Large test tube (slightly melting), stand, boss, clamp, Pasteur pipette, safety glasses, protective gloves.

Chemicals Bromine, aluminum welding wire, sandpaper.

Attention! This experiment must be carried out in a well-ventilated fume hood. Bromine is corrosive and very toxic when inhaled. Safety glasses and protective gloves must be used at all times.

Spectacular Chemical Experiments. Herbert W. Roesky
Copyright © 2007 WILEY-VCH Verlag GmbH & Co. KGaA, Weinheim
ISBN: 978-3-527-31865-0

Experiment 61 Sparks and Shining Fire

Experimental Procedure Using the Pasteur pipette, 1–2 g of bromine are placed into the test tube, which is fixed on the stand. An aluminum welding wire, one end of which has been roughened with the sandpaper, is placed into the bromine. After a short time the reaction occurs, producing bromine vapors and (sometimes) sparks and flames.

The aluminum wire should be 30 cm long, although aluminum foil can also be used for this demonstration.

Explanation Bromine reacts with aluminum at room temperature in an exotherm reaction to give $AlBr_3$:

$$3/2\ Br_2 + Al \rightarrow AlBr_3\ \Delta H_x° -527\ kJ\ mol^{-1}$$

Waste Disposal The remaining bromine in the test tube evaporates due to the strong reaction heat. If some elemental bromine is left, sodium thiosulfate is added; this reduces bromine to bromide. The residues in the test tube are hydrolyzed carefully with water and then flushed down the drain.

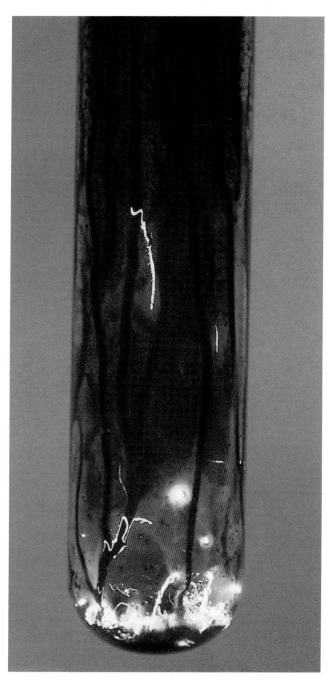

Experiment 61: Bromine and aluminum wire.

Experiment 62
Like Magic... the Reduction of Copper Oxide

Insomuch, that if thy former things were small, thy latter things would be multiplied exceedingly. For inquire of the former generation, and search diligently into the memory of the fathers (for we are but of yesterday, and are ignorant that our days upon earth are but a shadow), and they shall teach thee: they shall speak to thee, and utter words out of their hearts.

Job 8, 7–10

Apparatus One glass tube (20–25 cm long, 1.5 cm diameter, slightly meltable) stand, boss, clamp, Bunsen burner, one stopper with a hole, one stopper with a bent glass tube supported by a capillary, magnesia boat, test tubes, PVC tube, safety glasses, protective gloves.

Chemicals CuO powder, H_2 from a cylinder.

Attention! Hydrogen and air form explosive mixtures! Safety glasses and protective gloves must be used at all times.

Experimental Procedure The glass tube is fixed horizontally on the stand. One end is closed with the drilled stopper and connected to the H_2 cylinder with a tube. The magnesia boat is loaded with the CuO powder (black) and placed in the center of the glass tube. The other side of the glass tube is closed with the stopper containing the bent glass tube, and hydrogen is passed carefully through the whole apparatus. After an oxihydrogen test (twice!), the hydrogen can be ignited at the end of the capillary. The magnesia boat is then heated with a Bunsen burner. After a short time the CuO reacts (glows), and red copper is produced. The oxygen combines with the hydrogen to produce water, which condenses at the end of the glass tube. During the reduction of CuO with H_2 the flame, which normally is invisible, turns red-yellow in color.

The hydrogen stream should be turned off only when the tube has cooled down to room temperature. Otherwise, an explosive oxihydrogen mixture might occur due to air entering the system.

Explanation Copper oxide reacts with hydrogen to give Cu and H_2O.

Spectacular Chemical Experiments. Herbert W. Roesky
Copyright © 2007 WILEY-VCH Verlag GmbH & Co. KGaA, Weinheim
ISBN: 978-3-527-31865-0

Experiment 62 Like Magic... the Reduction of Copper Oxide

$$CuO + H_2 \rightarrow Cu + H_2O$$

This reaction shows very well that copper and oxygen react in a constant relation to produce copper(II) oxide (the law of constant proportions). If the experiment is carried out with different amounts of copper oxide, the following result is obtained:

Content 2.000 g – Yield 1.595 g
Content 4.000 g – Yield 3.159 g

In both of these cases the mass relationship of copper to oxygen is $4:1$.

Waste Disposal

The resulting copper is placed into a container for collecting copper.

Experiment 63
Electric Current from a Beer Can

These days, when fruits are expected to grow before the blossom, and therefore quite a few things are disdained because they do not heal immediately all the wounds, fertilize the fields, or turn the millwheel, many a person forgets that science has an inner target, and lose the true interest in searching knowledge, just knowledge.

Alexander von Humboldt

Apparatus	Aluminum beer can, 2-V motor, cable, stand, boss, clamp, sandpaper, two alligator clamps, graphite electrode, voltmeter, safety glasses, protective gloves.
Chemicals	2 M NaCl solution, 2 M KNO_3 solution.
Attention!	Safety glasses and protective gloves must be used at all times.
Experimental Procedure	First, the entire cover of the can is cut off, and the inside is roughened with the sandpaper. The 2 M NaCl solution is then filled into the can. Using the cables and a clamp, a small electric motor is connected to the edge of the can and to a carbon electrode. When the carbon electrode is dipped into the NaCl solution the small motor turns very quickly. If a voltmeter is connected instead of the motor, it shows a voltage of 0.4 to 1 V.
Explanation	This experiment cannot be carried out with a 2 M KNO_3 solution. Only the chloride ions are able to react with the cations of the aluminum layer under complex formation.
Waste Disposal	After completing the experiment, the aluminum/sodium chloride solution is flushed down the drain.
Reference	M. Ducci, M. Oetken, *Praxis der Naturwissenschaften Chemie* **2000**, *49*, 1.

Experiment 64
Magnesium Powder Burning in the Air

"How strangely does the dawnlike, murky light
seep through the trees and bushes.
How it pries and even penetrates
into ravines and gaping chasma.
Here fumes arise, here vapors hover,
a fire glows from mists below;
now it flickers like a tender thread,
now it gushes in a bursting spring.
Here it winds a crooked path
through the valley in a hundred veins;
there it crowds into a corner,
only sparkling now and then.
Suddenly there is a geyser
of sparks like incandescent grains of sand.
And look! The mountain wall from top to bottom
ignites and seems on fire."

Johann Wolfgang von Goethe, *Faust II*

Apparatus	Bunsen burner, large fire-resistant support, safety glasses, protective gloves.
Chemicals	Magnesium powder, magnesium ribbon (10 cm long).
Attention!	The reaction is strongly exothermic. A very bright flame is produced. The use of sunglasses and leather gloves is strongly recommended.
Experimental Procedure	A cone-shaped heap of magnesium powder (total weight 12 g) is placed on the fire-resistant support. One end of the magnesium ribbon is placed on top of the cone; the other end is ignited with the Bunsen burner.
Attention!	A very hot and bright flame is produced.
Explanation	After cooling the products to room temperature, a difference in composition between an outer layer (MgO) and an inner layer (Mg_3N_2)

Spectacular Chemical Experiments. Herbert W. Roesky
Copyright © 2007 WILEY-VCH Verlag GmbH & Co. KGaA, Weinheim
ISBN: 978-3-527-31865-0

can be noticed. The reaction occurs according to the following equation:

$$14\ Mg + O_2 + 4\ N_2 \rightarrow 2\ MgO + 4\ Mg_3N_2$$

In order to prove the presence of magnesium nitride, water is poured onto the cold product. A small amount of ammonia is formed, which turns litmus paper from red to blue.

$$Mg_3N_2 + 6\ H_2O + 3\ Mg(OH)_2 + 2\ NH_3$$

In order to increase the amount of Mg_3N_2 produced during the combustion, a small nitrogen stream from a cylinder is passed over the burning cone. WARNING: the nitrogen stream must be very gentle – otherwise, burning magnesium particles will be blown about and these will cause fire or severe burns! Under no circumstances should the nitrogen stream be directed towards the spectators.

Waste Disposal

After hydrolysis, the residues can be deposited in the domestic garbage.

Experiment 65
The Alchemist's Gold

The famous alchemist Nicolas Flamel (1330–1418) describes how he produced gold:

"This happened on a Monday, January 17th, around noon, in the Year of Grace 1382, in my house, with only my wife Perenella being present ... Later on, paying attention to every single word in my book, I let the red stone fall on almost the same amount of mercury, again in the presence of Perenella, in the same house at about 5 o'clock in the afternoon; I converted it into almost the same amount of pure gold, which was certainly much smoother than ordinary gold. This is the truth: I repeated the procedure three times with the help of Perenella, who also knew how to do it because she was assisting me; without doubt, she too would have been successful, if she had conducted the experiment by herself. Indeed, I had achieved much before this success for the first time; but nonetheless it was a great pleasure to see the fantastic works of nature inside vessels, and to reflect on this... For a long time I feared that Perenella would not be able to keep her exuberant, and would tell her relatives what treasures we own. For great joy just as great grief drive one out of ones mind; but thanks to the grace of the Almighty, I received not only one great blessing, he also gave me a virtuous and prudent wife; for not only was she intelligent, she was also sensible and more discrete than women usually are. Above all, she was very religious, and because she was already relatively old and did not expect to have any children, she began, like me to think about God and to devote herself to charity."

Apparatus A 1-L aluminum pot with a lid, Bunsen burner, tripod, lighter, safety glasses, protective gloves.

Chemicals Black gunpowder, pyrite crystal or a piece of brass.

Attention! Deflagration and smoke will occur. Keep away from the experiment at a distance of about 1 meter. SO_2 is toxic when inhaled. Safety glasses and protective gloves must be worn at all times.

Experimental Procedure Several hours before carrying out the experiment, the pyrite crystal is fixed under the lid with molten wax. Before placing the black gunpowder (maximum 3 g) into the pot, it should be made clear to the

Spectacular Chemical Experiments. Herbert W. Roesky
Copyright © 2007 WILEY-VCH Verlag GmbH & Co. KGaA, Weinheim
ISBN: 978-3-527-31865-0

spectators that the pot is empty. The lid is lifted carefully so that nobody sees the crystal pyrite. The black gunpowder is filled into the pot, after which the pot is closed with the lid; it is then heated up with the Bunsen burner. After a short time the black gunpowder decomposes under deflagration and smoke. Usually, the wax melts and the crystal falls to the bottom of the pot. If this cannot be heard, the lid must be removed horizontally over the edge of the pot.

Explanation

Black gunpowder contains sulfur, potassium nitrate, and charcoal. When it is heated it decomposes to produce KNO_2 and oxygen; the latter oxidizes the sulfur and the charcoal, producing intense heat.

Experiment 66
Imitate a Spider

Arachna, princess of Kolophin in Lydia, where crimson manufacturing was famous, was a very skilful weaver, so that even Athena could not compete with her. Once Athena saw a fabric which was woven by Arachna representing the Olympic love stories. The goddess tried to find a defect. When she did not find any she became very furious, ripped the fabric and plotted revenge. Arachna was scared and tried to escape climbing down from a balcony. At this moment Athena transformed her into a spider – an insect which she hated most – and the rope was transformed into a spider's web where the frightened Arachna was hiding.

Robert von Ranke-Graves, *Griechische Mythologie Quellen und Deutung* (Reinbek bei Hamburg 1968)

It was summer. Already a silvery thread was moving in the wind. I sat down on one of those close to a small wrinkled spider who immediately after she saw me began to cheer me with the story of her life.
Once, she began to talk nostalgically, about 2000 years ago, she was a very beautiful valkyrie, riding on a proud horse and beloved by every man. When she became old, nobody cared any more about her, only the devil. So she became a witch, she was riding on a broom with malicious intentions. But when she had celebrated her thousandths birthday, even the devil did not remain true. Then she took the pot with the ointment and transformed herself into a spider. She constructed this kind of airship, and is now always sailing with the wind. And when people are saying: Old women's summer! Then she does not care. Apa.

Wilhelm Busch, Eduards Traum (in: *Teufelsträume. Phantastische Geschichten des 19. Jahrhunderts*. Ed. Horst Heidtmann, München 1983)

Apparatus	Two 250-mL beakers, 250-mL measuring cylinder, 2-mL Pasteur pipette, two glass rods, moving coil, safety glasses, protective gloves.
Chemicals	1,6-Diaminohexane, sebacic acid dichloride (decanodioic acid dichloride), hexane, H_2O.
Attention!	1,6-Diaminohexane and hexane are toxic when inhaled and on contact with the skin. Safety glasses and protective gloves must be worn at all times.

Spectacular Chemical Experiments. Herbert W. Roesky
Copyright © 2007 WILEY-VCH Verlag GmbH & Co. KGaA, Weinheim
ISBN: 978-3-527-31865-0

Experiment 66 Imitate a Spider

Experimental Procedure

2.2 g of 1,6-diaminohexane are placed in a beaker and dissolved in 50 mL of water. Then, 1.5 g sebacic acid dichloride and 50 mL hexane are filled into the second beaker. This solution is now carefully poured into the aqueous phase of the first beaker. A cloudy film is formed at the interphase of the aqueous and organic phases. Using the glass rod, it is possible to form a thread; this must be done very carefully so that the nylon film between the organic and aqueous phases is not disturbed. As it is produced, the nylon thread can be rolled up with a moving coil.

In order to demonstrate the boundary between the organic and aqueous phases better, the aqueous phase may be colored with some sodium hydroxide and phenolphthalein.

Explanation

Polyamides are formed by the polycondensation of dicarbonic acids and diamines. In the presence of sodium hydroxide, the decanodioic acid dichloride ($C_{10}H_{16}Cl_2O_2$) reacts with 1,6-diaminohexane to give nylon (6,10) under the elimination of sodium chloride. The numbers in brackets indicate the number of carbon atoms of the diamine and the dicarbonic acid, respectively.

Waste Disposal

The organic residues are collected in the container used for halogen-containing compounds.

Reference

W. Amann, *Elemente Chemie II*, Stuttgart, 1989.

Experiment 67
Is it Methyl Alcohol or Ethyl Alcohol ? (A Simple Test with Boric Acid)

Liebig reports:

"At the market place in Darmstadt I picked up from a street trader how he transforms detonating silver to his toy torpedos. When he dissolved his silver the first red vapors showed me that he used nitric acid and a liquid which he used to clean the dirty jacket collars and which smelled like alcohol . . ."

Apparatus — Two porcelain dishes (8–10 cm in diameter), two 10-mL measuring cylinders, lighter, safety glasses, protective gloves.

Chemicals — Methyl alcohol, boric acid, ethyl alcohol.

Attention! — Methyl alcohol and ethyl alcohol are flammable. Especially methyl alcohol is toxic, it should not be inhaled. Safety glasses and protective gloves must be used at all times.

Experimental Procedure — About 1 g of boric acid is placed into each of the porcelain dishes. Then 10 mL of methyl alcohol are filled into the first porcelain dish, and 10 mL of ethyl alcohol into the second dish. When the alcohols are ignited, the flames of both the alcohols show different colors. The methyl alcohol flame is green; the ethyl alcohol flame is ordinary yellow. Only before the ethyl alcohol flame extinguishes, it turns into light green, because particles of the boric acid are carried along.

Explanation — In the presence of sulfuric acid and alcohols boric acid reacts to give boric esters which are slightly volatile and produce the green color of the flame. It is known that the reaction to form boric acid methyl ester occurs more completely than the corresponding reaction of the ethyl alcohol. Under the above mentioned conditions without sulfuric acid only methyl alcohol reacts to give the ester.

$$B(OH)_3 + 3\ CH_3OH \rightarrow B(OCH_3)_3 + 3\ H_2O$$

Spectacular Chemical Experiments. Herbert W. Roesky
Copyright © 2007 WILEY-VCH Verlag GmbH & Co. KGaA, Weinheim
ISBN: 978-3-527-31865-0

Experiment 67 Is it Methyl Alcohol or Ethyl Alcohol ? (A Simple Test with Boric Acid)

This is an easy method to distinguish methyl alcohol from ethyl alcohol.

Waste Disposal   The residues of boric acid are collected in the container used for less toxic residues.

Experiment 68
Oxygen Content of the Air

Scheele, Priestley and Lavoisier could prove that the air is a mixture of two gases: one part which maintains the combustion (oxygen), and another part which does not maintain the combustion (nitrogen).

Apparatus

A 2-L glass bell jar out of white glass with an open end (wall 5–7 mm thick), rubber stopper, cork ring, small porcelain dish, Bunsen burner, glass rod (50 cm long), pneumatic glass trough, safety glasses, protective gloves.

Chemicals

White phosphorus, $CuSO_4$, water.

Attention!

White phosphorus is extremely toxic and spontaneously inflammable in air. The experiment should be performed in a well-ventilated fume hood. Safety glasses and protective gloves must be used at all times.

Experimental Procedure

Before carrying out the experiment, the bell jar is closed with the rubber stopper, and then arranged with the open bottom at the top. Five portions of water (each of 300 mL) are filled into the bell jar, one after the other, and each time the upper level is labeled by an intensively colored graduation mark. Thus, a total of 1.5 L water is filled into the bell jar. The water and the stopper are then removed, such that the upper part of the bell jar is now labeled by graduation marks into five equal volumes.

In order to carry out the experiment, the labeled bell jar is placed vertically in the pneumatic glass trough. Water is then poured in up to the first mark. The cork ring and the porcelain dish are placed inside the bell jar, so that the porcelain dish is "swimming" on the cork ring. Shortly before performing the experiment, 0.5 g of white phosphorus is placed in the porcelain dish (for this purpose, the bell jar is removed from the water). After returning the bell jar to its previous position, the white phosphorus is ignited with the help of a glass rod which has been heated at one end and passed through the feed hole of the bell jar. After igniting the phosphorus, the glass rod is

Spectacular Chemical Experiments. Herbert W. Roesky
Copyright © 2007 WILEY-VCH Verlag GmbH & Co. KGaA, Weinheim
ISBN: 978-3-527-31865-0

removed immediately and the bell jar closed with the rubber stopper. An intense reaction, with formation of a luminous flame takes place. The white smoke is formed from phosphorus oxides and phosphorus oxoacids, which react exothermally with the water vapor. The reaction ends after about 1 minute.

When the reaction is over, the water inside the bell jar begins slowly to rise. Only after 10–15 minutes, when the bell jar has reached room temperature, is it possible to determine the amount of used oxygen due to the difference in water levels inside and outside the bell jar.

Explanation

The complete oxidation of P_4 with oxygen produces P_4O_{10}, which reacts exothermally with water, with the formation of orthophosphoric acid (H_3PO_4).

$$P_4 + 5\,O_2 \rightarrow P_4O_{10}$$
$$P_4O_{10} + 6\,H_2O \rightarrow 4\,H_3PO_4$$

Waste Disposal

Phosphorus residues are converted with 1 M copper sulfate solution into copper phosphide, which is oxidized with a strongly alkaline sodium hypochlorite solution. By adding milk of lime, this is converted into calcium phosphate and copper hydroxide. Both solids are slightly soluble in water and may be disposed of in the container for collecting less-toxic inorganic substances. The aqueous solution is neutralized with sulfuric acid and flushed down the drain.

Experiment 69
Rapid Rust

Chemistry not only creates spiritual order, it is adventure and aesthetic experience.

Cyril Hinshelwood

Apparatus	A 1-L round-bottomed flask without ground joint, 400-mL beaker, one glass tube (50 cm long, 9 mm outer diameter), one rubber stopper with a hole (for the glass tube), stand, bosses, clamps, safety glasses, protective gloves.
Chemicals	Steel wool, water, fluorescein sodium salt, 6 M hydrochloric acid.
Attention!	6 M HCl must be always used in a well-ventilated fume hood. Safety glasses and protective gloves must be used at all times.
Experimental Procedure	The round-bottomed flask is filled with steel wool (ca. 45 g), followed by the addition of 6 M hydrochloric acid (ca. 250 mL). After a few minutes, the HCl solution is removed and the flask is cleaned with tap water and attached to the stand with the open end pointing downwards. Now, the rubber stopper with the glass tube is attached. The lower end of the glass tube is placed in the beaker with the colored water (300 mL). After about 10–15 minutes, the colored water has passed through the glass tube into the flask, and for 30 minutes it continues to drip into the flask.
Explanation	The experiment is successful only when the steel wool is treated with the hydrochloric acid immediately before the demonstration; this ensures a largely oxide-free surface. Corrosion of the steel wool begins immediately after the cleaning process. The oxygen of the air inside the flask is used to oxidize the iron, which causes the reduction in pressure. This is shown by the rise in the level of colored water.
Waste Disposal	The hydrochloric acid is neutralized with sodium hydroxide and flushed down the drain. The steel wool can be disposed of in the domestic garbage.

Spectacular Chemical Experiments. Herbert W. Roesky
Copyright © 2007 WILEY-VCH Verlag GmbH & Co. KGaA, Weinheim
ISBN: 978-3-527-31865-0

Experiment 70
Shining Dry Ice

The physicist Georg Christoph Lichtenberg made experiments with fixed air (CO_2) and reported:

"It did not rise (a balloon), but it sank so slowly that I thought it could stay on a somehow heavier air. Fortunately a big bottle filled with fixed air was close by . . .; it was filled into another container, when I throw then the small balloon into this container it is floating without touching the walls of the container . . . (We) have here a freely floating body which rises when it is pushed downwards and sinks when it is raised."

"In twenty years you will be more disappointed about the things you did not do than about those you did. Therefore weigh the anchor and sail away out of the safe harbor. Catch the trade-winds with your sails. Discover your dreams."

Mark Twain

Apparatus — A large fire-resistant support, Bunsen burner, safety glasses, protective gloves.

Chemicals — Magnesium ribbon, CO_2 block (ca. $14 \times 12 \times 5\,cm^3$), magnesium powder.

Attention! — It is dangerous to look at the bright flame of burning magnesium. Safety glasses and protective gloves must be used at all times.

Experimental Procedure — A hole (2 cm deep, 3 cm diameter) is drilled into the CO_2 block. 10 g of magnesium powder are filled into the hole, and the magnesium ribbon (10 cm long) is placed in the magnesium powder. The magnesium ribbon is ignited with the Bunsen burner. The burning ribbon ignites the magnesium powder. After a few seconds the magnesium powder reacts and the CO_2 block begins to shine with a bright flame.

Explanation — The burning magnesium powder reacts with oxygen to give MgO, producing a bright flame. Inside the hole, MgO is formed according to the following equation:

Spectacular Chemical Experiments. Herbert W. Roesky
Copyright © 2007 WILEY-VCH Verlag GmbH & Co. KGaA, Weinheim
ISBN: 978-3-527-31865-0

$$Mg + CO_2 \rightarrow MgO + CO$$

Due to the high temperature and the formation of gaseous carbon dioxide, most of the MgO is ejected from the hole. Thus, all inflammable liquids or solids nearby should be removed before the experiment is started.

Waste Disposal The resulting MgO is disposed of in the domestic garbage.

Experiment 70 Shining Dry Ice | 175

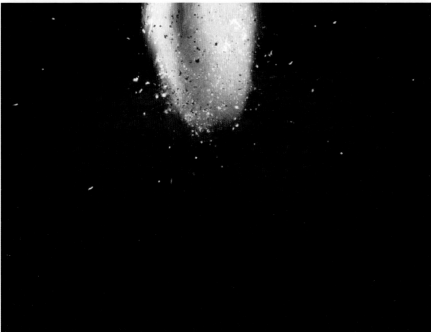

Experiment 70: On top: the block of dry ice and magnesium powder. At the bottom: during the reaction.

Experiment 71
Smoke Rings

The meeting of two personalities is like the contact of two chemical substances. If a reaction occurs, both of them are transformed.

Carl Gustaf Jung

Teach me the art to make small steps.

Antoine de Saint Exupéry

Apparatus Two gas wash bottles, two Y-shaped glass tube connections, one small drying tube (straight, about 10 cm long, about 2 cm in diameter), which is also called calcium dichloride drying tube, one half-blower, stand, boss, clamps, tubing, safety glasses, protective gloves.

Chemicals Concentrated hydrochloric acid, concentrated ammonia solution.

Attention! Ammonia solutions are irritating to the skin and eyes, and the vapors must not be inhaled. Concentrated hydrochloric acid is very corrosive; contact with the skin and eyes must be avoided. All procedures must be carried out in a well-ventilated fume hood. Safety glasses and protective gloves must be used at all times.

Experimental Procedure For this experimental procedure, both gas wash bottles are attached to the stand. The Y-shaped tube connections with short tubes are fixed first on both sides of the wash bottles. The half-blower is connected to the free part of one of the Y-shaped tubes; the drying tube is attached to the other Y-shaped part. For a better demonstration, the drying tube is fixed vertically to the stand. It is important that the gas inlet tubes of the wash bottles end about 1 cm above the bottom of the bottle.

For the demonstration, 5 mL of concentrated hydrochloric acid is placed into the first wash bottle, and 5 mL of concentrated ammonia solution is poured into the second wash bottle. The gas inlet tubes

Spectacular Chemical Experiments. Herbert W. Roesky
Copyright © 2007 WILEY-VCH Verlag GmbH & Co. KGaA, Weinheim
ISBN: 978-3-527-31865-0

should not dip into the solutions. By pressing the half-blower, both gases will react in the gas phase to produce white ammonium chloride, which is released through the drying tube (a type of chimney or "smoker's mouth"). Under these conditions, the ammonium chloride "smoke" is released in the form of rings, resembling a smoker who forms the typical smoke rings with their mouth.

This demonstration is especially impressive when the tube is painted brown and the wash bottles are hidden behind a "Paper Indian", who is smoking his peace-pipe.

Explanation

Gaseous HCl and NH_3 react to produce white NH_4Cl smoke.

Waste Disposal

The residues in the wash bottles are diluted with water and flushed down the drain.

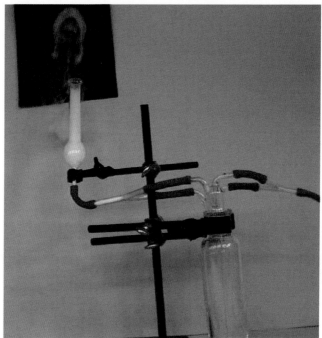

Experiment 71: On top: the experimental set-up. At the bottom: a smoke ring in front of a black background.

Experiment 72
Saturn's Rings

In 1799, the writer Joseph Rieckert reported ridiculed experimental lectures:

"Presently, people in Weimar only talk about gas, oxigena, combustible materials, light and heavy liquid things.
All the women and men of Weimar apparently want to become chemists, and the whole city seems to turn into a melting furnace."

However, due to experimental failures, the lectures were cancelled.

"... finally some experiments went off very badly, so that many participants ended up with burnt faces and dresses, and went home very angry."

Apparatus	A 1-L beaker, test tube (160 × 16 mm, slightly melting), test tube, stand, Bunsen burner, safety glasses, protective gloves.
Chemicals	Candle wax, tap water.
Attention!	Hot candle wax can cause severe burns to the skin. Safety glasses and protective gloves must be used at all times.
Experimental Procedure	About 1 g of candle wax is placed in the slightly melting test tube. The test tube, when fixed to the stand, is heated with the Bunsen burner. Heating is continued until the smoke produced above the melt almost reaches the open end of the test tube. Now, the test tube is immediately dipped into the beaker filled with water. The wax escapes from the test tube explosively as a smoky cloud, and it ignites at about 1 m above the table to form a burning fiery ring.
Explanation	It is important to heat the test tube for long enough; otherwise, the wax only escapes as a smoky cloud but does not ignite. Pyrolysis causes wax reactive products to be formed, and these ignite spontaneously in air due to their very small particle size.

Spectacular Chemical Experiments. Herbert W. Roesky
Copyright © 2007 WILEY-VCH Verlag GmbH & Co. KGaA, Weinheim
ISBN: 978-3-527-31865-0

Experiment 73
Oxygen from Ag_2O

Cognition means to combine exterior perception and interior ideas, and to value their harmony.

J. Kepler (1619)

Apparatus	Slightly melting test tube (160 × 15 mm), Bunsen burner, stand, boss, clamp, wooden splint, safety glasses, protective gloves.
Chemicals	Ag_2O.
Attention!	Safety glasses and protective gloves must be used at all times.
Experimental Procedure	The slightly melting test tube is fixed on the stand and filled with about 1 g of Ag_2O. The Ag_2O is strongly heated with a Bunsen burner. After some time, the Ag_2O decomposes to generate oxygen which can be detected with a glowing wooden splint, which starts to burn. After further heating, the oxygen formation ceases and a silvery solid metallic residue is formed.
Explanation	When heated, Ag_2O decomposes to yield silver and oxygen. Unlike HgO, which usually serves to produce oxygen, no volatile toxic metal is formed with Ag_2O.
Waste Disposal	The silver residues are separately disposed of in the appropriate container, and can be re-used after purification.

Experiment 74
Flour Dust Explosion

The greatest things in the world are brought about by other things which we count as nothing: little causes we overlook but which at length accumulate.

Georg Christoph Lichtenberg

Flour Dust explodes: An Inferno: Mill in ruins – 14 Persons Presumed Dead Bremen (dpa/ap) A severe flour dust explosion in a Bremen mill on Tuesday evening has presumably killed 14 people.
Bremen (dpa/ap) *A severe flour dust explosion in a Bremen mill on Tuesday evening has presumably killed 14 people.*
The explosion which officials said had the detonation power of at least 20 tons of explosives razed the Roland mill close to the Bremen wood and factory harbour to the ground. Four bodies had been recovered by Wednesday afternoon. Ten people are missing, and the chance of finding any survivors is negligible.
The missing workers, the caretaker and three of his relatives were probably killed by the tremendous shock wave produced by the explosion. During the press conference called by the Minister of Internal Affairs Fröhlich and attended by the Mayor, Koschnik, the speaker of the fire department said, that nobody within a 50 meter distance could have survived due to the enormous shock wave.
12 of the 17 rescued could leave the hospital after treatment in the out-patient department. Five severely injured are still in hospital.
The inferno began at 21.24 hours on Tuesday evening immediately after the shift change. In the Bremen wood and factory harbour a flour dust explosion occurred in a storehouse for forage cereal owned by the company Rolandmühle. Within a radius of 2 kilometers windows were shattered and stones and brickwork were thrust around.
Buildings and silos collapsed like houses of cards. Storehouses and factories burnt like wildfire. Flames were up to 40 meters high. A 40 meters high tower on top of the administration building collapsed. The shock wave was registered in the neighbouring communities.
The fire-fighting operations were extremely dangerous due to the presence of a nearby wool storage house. However, the fire could be extinguished before it spread to the storage house. A ship anchored nearby had to be moved to another location after the superstructure had been destroyed by falling stones. It took the fire department until the early morning to get the fire under control. The clearing-up and the search for persons will probably take several days. The material loss will be more than 50 millions of DM.

Spectacular Chemical Experiments. Herbert W. Roesky
Copyright © 2007 WILEY-VCH Verlag GmbH & Co. KGaA, Weinheim
ISBN: 978-3-527-31865-0

Experiment 74 Flour Dust Explosion

Apparatus — A Plexiglas box with an open top (ca. $20 \times 12 \times 18\,cm^3$; with a hole in one of the sides) and with a supply to a funnel which is fixed to a stand inside the box, small candle, lighter, filter paper, spatula, thin cardboard, glass tube (60 cm long, 8 mm diameter), gas burner, safety glasses, protective gloves. An illustration of the experimental set-up is shown below.

Chemicals — Cylinder with oxygen and reducing valve, lycopodium powder, flour.

Attention! — Safety glasses and protective gloves must be used at all times.

Experimental Procedure — Before carrying out the experiment, the lower part of the funnel is covered with a piece of paper, on which 0.5–1 g of lycopodium dust is placed. The funnel is connected via a tube to the oxygen cylinder. A fixed candle is placed inside the box, 10 cm away from the funnel, and ignited. The box is covered with the thin cardboard, and a strong intermittent oxygen stream is introduced. The lycopodium dust inside the box swirls and either explodes or deflagrates due to the ignition with the candle. The cardboard is thrown upwards and the shock wave extinguishes the candle.

This experiment can be compared to a flour dust explosion in a mill where, very often, such an explosion is due to an electric discharge.

A simple experiment to show a flour dust explosion can be made by using a 60 cm-long glass tube (8 mm diameter). An 8-cm length of one end of the glass tube is filled with lycopodium dust or very fine dry flour (which usually is dried in an oven). The tube should not be filled completely, as this allows the lycopodium dust (flour) to spread about when it is blown out. The gas flame of the gas burner is switched on, and the dust in the glass tube is blown into the flame. The well-spread lycopodium dust burns, producing a very large flame.

The glass tube should not be directed towards the spectators. In addition, the demonstrator should blow into the flame from the front, so that the flame does not endanger the spectators.

Experiment 74: Flour dust explosion; the experimental set-up.

Experiment 75
Bromine and Potassium

Because of the eternal flame burning inside it seems to be red.

Dante, *Inferno*

Bromine and mercury are the only elements which are liquid under normal conditions. Bromine is a deeply brown-red liquid, and at room temperature has a high vapor pressure. These vapors are very aggressive; bromine irritates the mucous membranes, with the reaction resembling that of a "cold". Bromine causes festering wounds which do not heal very well. In the vaporous state, 1 L of bromine weighs 7.1 g; thus, it is five-and-a-half-times heavier than air.

Apparatus	A thick-walled test tube, stand, clamp, boss, iron wire (40 m long, 1 mm diameter), safety glasses, protective gloves.
Chemicals	Bromine, potassium.
Attention!	This experiment must be carried out in a well-ventilated fume hood. Bromine is extremely toxic, and the inhalation of vapor and skin contact should be avoided. Potassium reacts explosively with water. Safety glasses and protective gloves must be used at all times.
Experimental Procedure	The test tube is fixed vertically with the clamp to the stand. 1 mL of bromine is filled into the test tube. The iron wire is bent in the middle, forming a right angle. A pea-sized piece of potassium is fixed to one end of the iron wire. When the potassium is introduced with the wire into the liquid phase, it reacts explosively with bromine to produce KBr.
Waste Disposal	The excess bromine is vaporized in the fume hood. The residue is reacted with alcohol and flushed down the drain.

Spectacular Chemical Experiments. Herbert W. Roesky
Copyright © 2007 WILEY-VCH Verlag GmbH & Co. KGaA, Weinheim
ISBN: 978-3-527-31865-0

Experiment 76
Current-Free Shining Flat-Bottomed Cylinder

Art is beautiful, but it is a lot of work.

Karl Valentin

Apparatus One flat-bottomed cylinder (30 cm high, 4 cm diameter), one stopper, scissors, one filter paper strip (25 cm long, 2.5 cm wide), Pasteur pipette with cap, pair of tweezers, safety glasses, protective gloves.

Chemicals Tetrakis(dimethylamino)ethane (TDAE).

Attention! The reaction should be carried out in a well-ventilated fume hood. Safety glasses and protective gloves must be used at all times.

Experimental Procedure The filter paper strip is soaked with TDAE, using the Pasteur pipette, and then placed in the flat-bottomed cylinder. The TDAE reacts with oxygen, showing a green-yellow chemiluminescence. In a darkened room the luminosity can be seen for about 10 minutes.

Explanation The reaction of TDAE with oxygen produces excited bisdimethylaminoketone which, by showing a green-yellow chemiluminescence, passes over into the ground state.

$$\begin{array}{c}Me_2N\\Me_2N\end{array}\!\!C\!=\!C\!\begin{array}{c}NMe_2\\NMe_2\end{array} + O_2 \longrightarrow \begin{array}{c}Me_2N\\Me_2N\end{array}\!\!C\!=\!O + \left[O\!=\!C\!\begin{array}{c}NMe_2\\NMe_2\end{array}\right]^*$$

Waste Disposal The paper strip can be disposed of in the domestic garbage.

References
- D. Woehrle, M.W. Tausch, W.D. Stohrer, *Photochemie: Konzepte, Methoden, Experimente*, Wiley-VCH, Weinheim. Internetseite www.theochem.uni-duisburg.de
- H. Bock, H. Borrmann, Z. Harlas, H. Oberhammer, K. Ruppert, A. Simon, *Angew. Chem.* **1991**, *103*, 1733.

Spectacular Chemical Experiments. Herbert W. Roesky
Copyright © 2007 WILEY-VCH Verlag GmbH & Co. KGaA, Weinheim
ISBN: 978-3-527-31865-0

Experiment 77
Rotating Advertising Column

Tradition is a lantern. The stupid holds tight to it, for the wise it lights the way.

George Bernard Shaw

Apparatus	One 250-mL beaker, brush, unbleached paper, glass rod, UV lamp, safety glasses, protective gloves.
Chemicals	Detergent, water.
Attention!	Safety glasses and protective gloves must be used at all times.
Experimental Procedure	In the beaker, 100 mL of water are added to about 10 g of detergent. This solution is applied with the brush to the unbleached paper (either by writing letters or drawing a pattern). When the paper is dry it can be irradiated with the UV light, when the letters or the pattern can be seen very well. They are seen to shine in the UV light, whereas untreated areas of the paper do not shine. Pure detergents, when in solution, do reflect UV light. After drying, the pre-prepared paper is fixed on a paper roll. Before irradiation, the paper roll is mounted on a slowly moving rotating plate. This can be compared to a rotating advertising column.
Explanation	Optical color brighteners usually are organic compounds which are deposited on the surface of fabric or paper, and turn the ultraviolet imperceptible part of white light into the perceptible blue part. The blue coloring makes the fabric brilliant, so that it appears particularly clean.
Reference	M. Schallies, *Kunststoffe, Farbstoffe*, Waschmittel, Bamberg 1980, S. 63.

Experiment 78
S$_4$N$_4$ – A Pick-Me-Up

... You have certainly already heard of the explosion which occurred during my lecture. The wounds are healed up, and we became an enormously interesting person. Queen Marie who is like an angel, beautiful, charming and gracious, invited me one day after the second lecture where all the persons involved, even the injured ones, attended, to come to the castle with my wife Agnes. She wanted to know her personally and when we arrived she gave Agnes a wonderful silver tea-set. The reason why she gave it to her made it especially precious. I created for the queen a little world in a glass, and I hope that she and the small charming princes will enjoy having it; there are goldfishes, small river fishes, salamanders and snails.

Liebig (1853), in a letter to Friedrich Mohr

Apparatus Hammer, solid iron plate, horn spatula, safety glasses, protective gloves.

Chemicals S$_4$N$_4$.

Attention! S$_4$N$_4$ may disintegrate explosively when heated or under pressure! Use only small amounts of S$_4$N$_4$ (0.05 g). S$_4$N$_4$ should only be prepared by an experimental chemist in a well-ventilated fume hood. Safety glasses and protective gloves must be used at all times.

Experimental Procedure A few S$_4$N$_4$ crystals are taken out of a small plastic bottle using a horn spatula, and placed on the iron plate. When the crystals are hit with the hammer, an explosion occurs. The loudness of the bang produced depends upon the amount of S$_4$N$_4$ used.

Explanation S$_4$N$_4$ decomposes, yielding sulfur and N$_2$:

$$S_4N_4 \rightarrow 2\,N_2 + \tfrac{1}{2}S_8$$

The released sulfur produces a typical "gunpowder smoke". S$_4$N$_4$ can be kept for years, in readiness for use in experiments.

Spectacular Chemical Experiments. Herbert W. Roesky
Copyright © 2007 WILEY-VCH Verlag GmbH & Co. KGaA, Weinheim
ISBN: 978-3-527-31865-0

S_4N_4 can be easily made from SCl_2 or S_2Cl_2 and dry ammonia.

$$6\ SCl_2 + 16\ NH_3 \rightarrow S_4N_4 + 2\ S + 12\ NH_4Cl$$
$$6\ S_2Cl_2 + 16\ NH_3 \rightarrow S_4N_4 + 8\ S + 12\ NH_4Cl$$

If the formation of sulfur is to be avoided, the hypothetical "SCl_3" is produced from S_2Cl_2 and chlorine gas in a CCl_4 solution.

$$S_2Cl_2 + 2\ Cl_2 \rightarrow 2\ \text{"}SCl_3\text{"}$$

SCl_3 can also be formulated as $2\ SCl_2 + Cl_2$. Moreover, the composition $S:Cl = 1:3$ yields the highest amounts of S_4N_4.

$$4\ \text{"}SCl_3\text{"} + 16\ NH_3 \rightarrow S_4N_4 + 12\ NH_4Cl$$

References

The formation of S_4N_4 is reported by:
- M. Villena-Blanco, W.L. Jolly, *Inorg. Syntheses* **1967**, 9, 102.
- M. Becke-Goehring, *Inorg. Syntheses* **1960**, 6, 123.

Experiment 79
Thunderclap

For months the Opus Magnum's mewed
Him up in total solitude.
This learned man, so meek and mild,
Looks like a charcoal-burner: wild
Complexion, black from ear to nose,
Eyes reddened by all the fires he blows.
Moment by moment he craves and longs;
Music for him's the click of tongs.

Johann Wolfang von Goethe, *Faust II*

Apparatus Stand, clamp, boss, iron tube (1.20 m long) with one end closed, stand, test tube without brim, glass pipette, filter paper, pair of tweezers, safety glasses, protective gloves.

Chemicals Potassium, pentachloroethane.

Attention! Potassium reacts explosively with water! The piece of potassium must be cut carefully, especially if there is a K_2O_2 crust on the surface of the potassium. This crust disintegrates – often explosively – under the pressure from the knife! Pentachloroethane should not be inhaled. The general precautions for handling chlorinated hydrocarbons must be observed. Safety glasses and protective gloves must be used at all times.

Experimental Procedure A pea-sized piece of potassium is cut from a bar, and the petroleum ether is removed with a filter paper. At the same time, the potassium is formed (with help from the filter paper) into the shape of a small lens. Using the tweezers, this potassium lens is placed into the test tube. The test tube is fixed by its upper part to the stand, such that it is held vertically inside the iron tube. Then, 6–8 drops of pentachloroethane are added to the potassium in the test tube, using the pipette. After some seconds a "cracking" noise can be heard. The test tube support is then released, so that the test tube falls to the bottom of the iron tube. A great "thunderclap" is produced, when the potassium reacts with the pentachloroethane to produce potassium chloride.

Explanation

Pentachloroethane reacts with potassium in an intense exothermic reaction, forming KCl and carbon. DANGER: it is essential that chlorine-containing solvents are NOT dried with alkali metals, as this may produce very large explosions.

Experiment 80
A Heavyweight does not Stick to the Bottom

The progress of science corresponds to an increasing number of reliable road signs.
Wilhelm Ostwald

The mixing of two gases can be shown very well with bromine and air. Due to their numerous collisions, the molecules are subject to permanent directional changes. The molecular velocities are distributed statistically, and are described by the Maxwell–Boltzmann velocity distribution. If the number of molecules is plotted against the molecular velocity in a gas, a curve with a maximum arises. The appropriate velocity is the one which occurs most frequently. The velocity curve depends on the molecular mass of the molecules.

Apparatus	Flat-bottomed cylinder (25 cm high, 4–5 cm diameter), Pasteur pipette, safety glasses, protective gloves.
Chemicals	Bromine.
Attention!	Bromine is highly toxic and corrosive. These experiments must be conducted in a well-ventilated fume hood. Safety glasses and protective gloves must be used at all times.
Experimental Procedure	The bromine is placed carefully on the bottom of the cylinder, using a Pasteur pipette. The cylinder is then closed. After about 30 minutes, bromine vapors spread out over the whole cylinder.
Waste Disposal	The cylinder is opened in a well-ventilated fume hood, and the brown color of the bromine allowed to disperse.

Experiment 81
Icarus and the Sun

Almost everybody knows the story of the Greek mythology. In the workshop of Daedalus, several apprentices were working. One of them, Tales, proved to be a better craftsman than Daedalus, who was afraid of him. Finally, he killed Tales and escaped with his son Icarus to Kreta. There they were sent to prison. Daedalus was a very skilful craftsman. He made wings with feathers and wax in order to escape from the prison. However, he warned his son Icarus against flying to close to the sun. But he did not obey. The wax melted due to the heat of the sun, and Icarus crashed.

This story fits very well with the experiment of using a fire-catching paper aeroplane and a burning glass.

Apparatus One sheet of A4 white paper, two stands, two clamps, stand rod, iron trough, black marker pen, focusing glass lens, burning glass, thin wire, safety glasses, protective gloves.

Chemicals KNO_3.

Attention! KNO_3 is an oxidizing agent, and its contact with any flammable substances must be avoided. Any combustion is dangerous. Safety glasses and protective gloves must be used at all times.

Experimental Procedure The potassium nitrate is dissolved in water to prepare a concentrated solution. A piece of paper is dipped into the saturated KNO_3 solution. It should be thoroughly soaked; it is then drained and dried overnight at room temperature, away from any combustible materials. This dried KNO_3-coated paper is folded into a small paper aeroplane.

The name "Icarus" is written on one side of the plane, using a marker pen. A couple of black dots are also placed randomly on the paper. The plane is suspended using a thin copper or aluminum wire on a clamp, and the whole assembly is brought to the sunlight. When the sun shines on the whole aeroplane for a few minutes, nothing happens. Then, a focusing lens is used, focusing especially on a point on the paper where nothing has been written. Still no change!

Spectacular Chemical Experiments. Herbert W. Roesky
Copyright © 2007 WILEY-VCH Verlag GmbH & Co. KGaA, Weinheim
ISBN: 978-3-527-31865-0

Then, the sunlight is focused on a point where "Icarus" is written, or a black dot appears on the paper. Combustion is initiated in a very few seconds. If the reaction is carried out without having written Icarus or placing the black dots on the aeroplane, nothing happens and the plane remains unchanged.

A control aeroplane is prepared in the same manner as the one above, except that it is made from a piece of paper that has not been treated with potassium nitrate. Exposing this to sunlight produces no evidence of burning on the white area, and only marginal, confined burning occurs on the black spots. For total burning of the paper, potassium nitrate and a blackbody are essential.

Explanation

As the sunlight is focused on the spot marked with ink, it functions as a blackbody and absorbs a large amount of energy. This energy is enough to initiate the oxidation of cellulose by potassium nitrate.

In physics, a blackbody is – in theory – a perfect absorber and emitter of invisible as well as visible radiation. Any substance, such as coal or pitch, that absorbs almost all of the light falling on it and reflects very little, approximates a blackbody.

Waste Disposal

The remaining KNO_3 solution is diluted with water and flushed down the drain.

Reference

M.G. Walawalkar, H.W. Roesky, *J. Chem. Educ.* **2001**, *78*, 912.

Experiment 82
Disposal of Sodium and Potassium Residues

It is true I cannot say whether things are going to change for the better, but what I do say is that things will never be right unless they do change.

Georg Christoph Lichtenberg

Apparatus A ceramic flower-pot (10 cm diameter), a large porcelain dish, pair of tweezers, safety glasses, protective gloves.

Chemicals Water, sodium and potassium residues.

Attention! Alkaline metals and water react intensely, producing fire. Safety glasses and protective gloves must be used at all times.

Disposing of sodium and potassium residues often causes accidents due to peroxides forming on the surface of alkaline metals when they are kept in the air for some time.

Sodium and potassium are often used for drying ether, tertiary amines, hydrocarbons, and aromatics. Any sodium and potassium residues must be disposed of, and usually this is done with 2-propanol. However, accidents occur when a peroxide containing alkali metal is pressed or cut, and when water containing alcohol or alcohol of low molar mass is used. Often, control of the reaction is lost when the alcohol is added too quickly.

Experimental Procedure The following method describes a less-expensive and safe procedure.
The bottom of a flower pot is covered inside with a filter paper, and half-filled with sand. About 1 g of an alkali metal residue is placed on top of the sand, using tweezers. The pot is then completely filled with sand, placed in the dish, and water poured into the dish to a depth of 2 cm. Due to the capillary action of the sand, the water rises in the pot – the presence of water is indicated by the dark color of the sand's surface. After one to two days, the water has decomposed

Experiment 82 Disposal of Sodium and Potassium Residues

the alkali metal, after which the sand can be washed, dried, and used again.

Explanation The reaction of sodium with the water is very slow and inaudible. The hydrolysis occurs within a certain time. This is due to the different surface quality of the metal pieces and for lack of determining the exact time when all the metal is consumed. Therefore, in the experimental procedure the time given for the hydrolysis includes a safety factor. This method avoids the use of combustible alcohols.

Waste Disposal The water used for washing the sand can be flushed down the drain.

Reference H.W. Roesky, *Inorg. Chem.* **2001**, *40*, 6855–6856.

Part VII
The Art Gallery of Chemistry

Every work of art is a child of one's time. Often it is the mother of our feelings. Every cultural period creates a special kind of art which cannot be reproduced. The effort to revive former art principles may only produce works of art which are doomed to failure.

Wassily Kandinsky

Color has come to every area of our life. How sad our villages and cities would be without the colorful walls of the houses. Goethe writes:

The colorful reflection is our life.

Thus, colorful chemical reactions are perhaps the most impressing experiences of our chemical education, and for this reason an emphasis must be placed on such reactions.

Small amounts of chemicals and coloring substances form the basic materials for not only colorful but also very impressive experiments. Moreover, by using a digital camera such compounds may be transferred into fascinating pictures, enabling the demonstrator to present his or her results to a wider public.

Experiment 83
Color Composition: Chemistry is Art

Two different kinds of principles exist:
The principle which shows the way to go and the principle which warms the heart. The first principle is science, the second one art. Both depend on each other, and none is more important than the other one. Without art science would be as useless as a pair of tweezers in the hand of a plumber. Without science art would be a mess of folklore and emotional charlatanism. The principle of art prevents that science becomes inhuman, the principle of science prevents that art becomes ridiculous.

Raymond Chandler, *Notebook: Big Thought*

Apparatus	Glass cuvette ($8 \times 6 \times 2\,cm^3$), dropping pipette, safety glasses, protective gloves.
Chemicals	Fluorescein, Congo red, indigo carmine, methylene blue, distilled water.
Attention!	Safety glasses and protective gloves must be used at all times.
Experimental Procedure	Differently colored indicators ($1\,mg\,mL^{-1}$ H_2O each) are added slowly, dropwise, to 80 mL of H_2O. The indicators are fluorescein, Congo red, methylene blue and indigo carmine. Depending on the amount of the added solution, and the speed of dropping, the resulting colors are more or less intense. The figure shows very clearly the color contrasts. This experiment shows the interplay of chemistry and art.

Experiment 83 Color Composition: Chemistry is Art

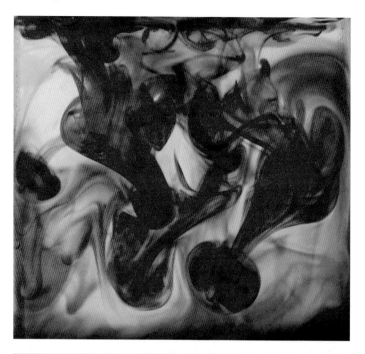

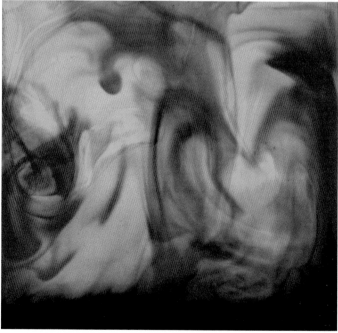

Experiment 83: Color composition.

Experiment 84
Underwater Dance

In 1911, Franz Marc and Wassily Kandinsky founded the artists' association "Blauer Reiter". Kandinsky describes it as follows:

We invented the name "Blauer Reiter" when we were drinking coffee in the small garden house. We both like the color blue, Marc likes horses, and I like horsemen. So the name came by itself.

Apparatus	Glass cuvette ($8 \times 6 \times 2\,cm^3$), pipette, safety glasses, protective gloves.
Chemicals	Methylene blue, distilled water.
Attention!	Safety glasses and protective gloves must be used at all times.
Experimental Procedure	Methylene blue ($1\,mg\,mL^{-1}$ H_2O) is added to 80 mL of water using a dropping pipette. The dye may be added dropwise or in a single aliquot (0.5–1 mL methylene blue). Depending on the amount of the added solution, the colors are more or less intense.
	Immediately after adding the methylene blue, about 10 photographs are taken with the digital camera, and the most beautiful are selected and printed on a color printer.
	In this way, chemical experiments can be presented to friends and relatives!
Waste Disposal	The content of the cuvette can be flushed down the drain.

Experiment 84 Underwater Dance

Experiment 84: Methylene blue in water.

Experiment 85
Blue Mist

For you, the younger ones, science will be much easier than it has been for the older ones. These days we are observing and making experiments, whereas in former times people were relying on assumptions, opinions even imaginations. For a long time those new ways were not clear, though one or the other may always have taken them.

Adalbert Stifter, *Nachsommer*

Apparatus	Glass cuvette ($8 \times 6 \times 2\,cm^3$), pipette, safety glasses, protective gloves.
Chemicals	$CuSO_4$, ammonia solution.
Attention!	$CuSO_4$ is toxic. Safety glasses and protective gloves must be used at all times.
Experimental Procedure	80 mL of a 0.1 M $CuSO_4$ solution are poured into the cuvette, after which an ammonia solution (25%) is added dropwise. A blue cloudy solution is seen to form immediately after the first drops are added.
Explanation	This experiment shows the light blue aquacomplex, the almost white (light blue) copper oxidehydrate as a precipitate, and the cornflower blue $[Cu(NH_3)_4]^{2+}$ cation.
Waste Disposal	The solution is poured into the container used for collecting heavy metal residues.

212 | *Experiment 85 Blue Mist*

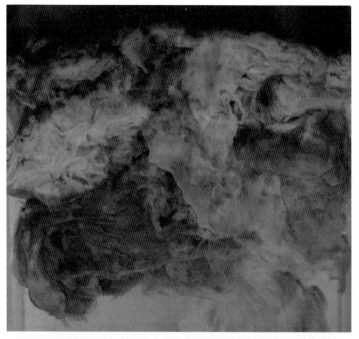

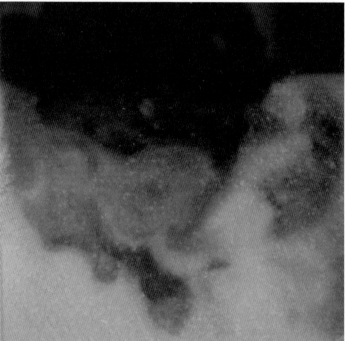

Experiment 85: Cu^{2+}_{aq} solution after the addition of varying amounts of aqueous ammonia solution.

Experiment 86
Colorful Clouds

Though I am not a scientist and cannot claim that I dealt with sciences, but I have read about these things, and I have tried my whole life to observe, and to think things over. Due to these efforts I have seen today the unambiguous signs that the clouds which remain during sun set, and from where some thunder could be heard, and which made you come up to me, will not bring any rain to this house or to some other place in this region. Maybe, they will spread when the sun is setting and will be scattered around. In the evening we will feel some wind, and certainly tomorrow will be again a beautiful day.

Adalbert Stifter, *Nachsommer*

Apparatus Glass cuvette ($8 \times 6 \times 2\,cm^3$), pipette, safety glasses, protective gloves.

Chemicals 0.1 M ammonium thiocyanate solution, distilled water, food coloring Ponceau 4R E 124 (blue) Brilliant Green E 142 (McCormick GmbH, Eschborn/Germany, or McCormick Inc., Baltimore, MD, USA), $AgNO_3$, distilled water, 0.1 M silver nitrate solution.

Attention! Safety glasses and protective gloves must be used at all times.

Experimental Procedure 80 mL of a 0.1 M ammonium thiocyanate solution are poured into the cuvette. To this is added 0.5 mL (1 mg mL^{-1} water) of the Ponceau 4R E 124 (blue) solution, and to the resultant solution is added dropwise a 0.1 M silver nitrate solution. Instead of a white precipitate (AgSCN), a colorful precipitate is formed, due to adsorption of the Ponceau 4R E 124 onto the surface of the AgSCN.

When Brilliant Green E 142 is used, a green-yellow precipitate from AgSCN is formed.

If a picture is to be taken without adding any coloring substance, the background of the cuvette may be altered by using differently colored papers.

Spectacular Chemical Experiments. Herbert W. Roesky
Copyright © 2007 WILEY-VCH Verlag GmbH & Co. KGaA, Weinheim
ISBN: 978-3-527-31865-0

Waste Disposal

The content of the cuvette is collected in the container used for silver residues.

References

- B.Z. Shakhashiri, *Chemical Demonstrations*, University of Wisconsin Press 1983.
- G. Harsch, H.H. Bussemas, *Bilder, die sich selber malen*, DuMont Buchverlag, Köln, 1985.
- F. Cherrier, *Chemie macht Spa<b>*, Verlag J.F. Schreiber GmbH, Esslingen.
- G. Wagner, *Chemie in faszinierenden Experimenten*, Aulis Verlag Deubner KG, Köln, 1980.
- F. Bukatsch, O.P. Krätz, G. Probeck, P.J. Schwankner, *So interessant ist Chemie*, Aulis Verlag Deubner KG, Köln, 1987.
- M. Tausch, M. von Wachtendonk, *Chemie SII Stoff-Formel-Umwelt*, Buchners Verlag, Bamberg 1993.
- S. Nick, J. Parchmann, R. Demuth, *Chemisches Feuerwerk*, Aulis Verlag Deubner KG, Köln, 2001.

Experiment 86 Colorful Clouds | 215

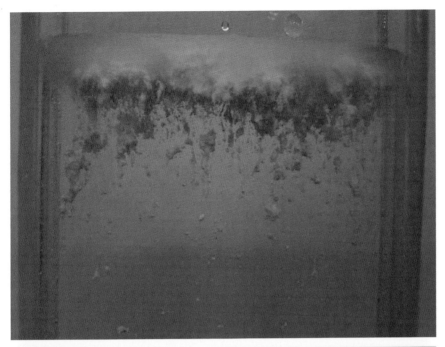

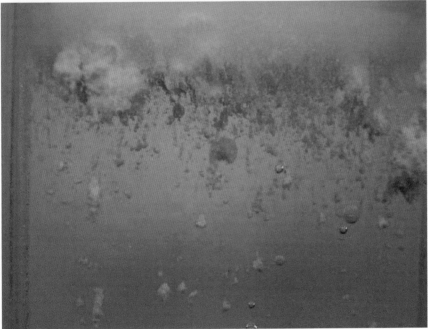

Experiment 86: On top: precipitation of AgSCN in the presence of Ponceau 4R E 124 (blue). At the bottom: precipitation of AgSCN in the presence of Brilliant Green E 142.

Conclusion

Over 160 years ago, Liebig first pointed out how important chemistry is for human beings, and even after so many years Liebig's words are still valid today. Nowadays, the virtual experiment on the computer still cannot be replaced by the experiment carried out in the laboratory.

"Without the exact studies of chemistry and physics physiology and medicine will not understand their most important duties of investigating the laws of life and the understanding and removal of abnormalities of the organism. Without knowledge of the chemical energy, the vital process cannot be investigated. Without knowledge in chemistry, a statesman stays away from the real life in his state, from its natural development and perfection; without this knowledge he cannot see things clearly enough, he will not be open minded enough to be aware of what is useful or harmful for the country and the people. The most important intrinsic interests, the increased and advisable maintenance and restoration of health are tightly connected to the study of science, especially chemistry; without any knowledge of nature's laws and nature's phenomena, the human mind fails in understanding the Creator's greatness and wisdom; for everything which fantasy or intelligence may imagine compared with reality is like a colored, shining empty soap bubble."

Justus von Liebig

In completing this book, I wish to cite Mark Twain:

"Do the right. Some people will be happy and all the others will be astonished."

Likewise, remember the wise advice of the king, when writing to the white rabbit:

"Begin at the beginning, and continue up to the end, then stop."

Alice's Adventures in Wonderland

Index

a
absorber   202
acetaldehyde   130
acidic acid butyl ester   119
activated carbon   71
activation energy   25
Adalbert Stifter   211, 213
Adolf von Bayer   61
Ag_2O 183
AgCl   133, 134
$AgNO_3$ 53, 109, 110, 111
AgSCN   213, 215
air   169, 171, 181
Albert Einstein   17, 95, 111, 125
Albert Schweizer   7
$AlBr_3$ 154
alchemist's gold   163
alcohol   203
alcohol flame   167
alcohol test   129
aldehyde   69
Alexander von Humboldt   159
Alice's adventures in wonderland   217
alkali metals   80, 203, 204
alkali silicates   95
aluminum   153, 154, 159
aluminum foil   154
aluminum wire   155
ammonia   40, 79, 122, 123, 196
ammonia solution   79, 121, 177, 211
ammonium chloride   178
ammonium molybdate   63
ammonium thiocyanate   213
aniline blue   71
anode   34, 56, 57, 58
Antoine de Saint Exupéry   177
aquacomplex   211
aqueous ammonia   212
aqueous phase   166

Aristotle   1
art   207
art gallery   205

b
$B(OCH_3)_3$ 167
$B(OH)_3$ 167
B.Z. Sakhashiri   XIV
$Ba(OH)_2 \cdot 8 H_2O$ 39
banana   149
beer can   159
bell jar   11
Belousov–Zhabotinsky reaction   138
Berzelius   47
bisdimethylaminoketone   191
blackbody   202
Blauer Reiter   209
bleach   47, 48, 78
bleaching   77, 87
blood cells   12
blue bottle   73
boiling water   23
bond energy   25
bones   2
boric acid   167
boric esters   167
brain   2
brass   163
brilliant green E 142   213, 215
bromine   153, 155, 189, 199
bromocresol green   119
bromothymol blue   139, 140

c
C.P. Snow   xiii
C_2H_5OH 129, 130
$CaCl_2$ 7, 8
$CaCl_2 \cdot 6H_2O$ 7, 8
$CaCO_3$ 44
calcium phosphate   28, 170

Spectacular Chemical Experiments. Herbert W. Roesky
Copyright © 2007 WILEY-VCH Verlag GmbH & Co. KGaA, Weinheim
ISBN: 978-3-527-31865-0

camera   xiv
candle wax   181
carbon electrode   159
Carl Gustaf Jung   177
cathode   34, 56, 57
cellophane   11
cellulose   202
CH₃OH   167
charcoal   164
charge-transfer   35
Charles R. Darwin   55
chemical education   205
chemical patterns   115
chemiluminescence   51, 52, 191
chemistry   207
chlorine   87
chlorophyll   49, 50
Christian Morgenstern   49
chromium(III)   129
clay cylinder   37, 38
cleaning liquid   77
ClO₂   27
CO   174
Co(NO₃)₂   67
CO₂   44, 45, 75, 80, 143, 173, 174
cobalt   67
cobalt salt pearl   68
colloids   93, 101
color   53, 59, 81, 207, 209
color brighteners   193
color composition   208
colored indicators   207
colorful   205
coloring substances   205, 213
combustion   66, 201
concentrated sulfuric acid   17
congo red   207
copper   158
copper(I) oxide   69, 70
copper(II) oxide   158
copper hydroxide   28
copper oxide   157, 158
copper sulfate   28, 69, 70, 170
Cr(III)   130
crystallization   41
Cu   211
Cu(II)   122
[Cu(NH₃)₄]²⁺   211
Cu(OH)₂   122
CuO   157, 158
CuSO₄   121, 169, 211
CuSO₄ · 5H₂O   61, 69, 137
Cyril Hinshelwood   171

d

D(+) glucose   69, 73, 103, 109
D₂O   25
Daedalus   201
dan clorox   47, 77
Dante   189
decanodioic acid dichloride   166
decomposition   17
density   25
depolarization   34
detergent   193
detonating gas   29, 30, 31
Dewar vessel   75
dextrose   73
1,6-diaminohexane   165, 166
digital camera   205
dinitrogen trioxide   75
dissipative structures   116
Dostojewski   73
drinking water   2
dry ice   79, 173, 175
dyestuff   91

e

E.F. Schwenk   135
Edgar Alan Poe   41
Edgar F. Smith   109
effusion   37
egg   43, 44, 45
electric discharge   107
electric motor   159
electrolysis   34, 55
electrolytic trough   34
electrolyzing   34
electronically activated   52
emitter   202
emitting light   52
entropy   40
Erasmus von Rotterdam   91
ether   19
ethyl acetate   49
ethyl alcohol   101, 115, 125, 125, 129, 167, 168
exothermic reaction   198
explosion   142, 185, 186
explosive mixtures   146
eyeball   2

f

Faust   v
FeCl₃   89, 90
[Fe(SCN)₃(H₂O)₃]   89
Fehling's solution   69, 70

Ferdinand Fischer   1
fire   27
flammable   201
flour   186
flour dust explosion   185, 187
fluorescein   11, 171, 207
fluorescence   50, 52
food colors   77
formaldehyde solution   99
Franz Kafka   xiv
Franz Marc   209
free enthalpy   40
freezing effect   39
fuel cell   33

g
galvanic battery   29
gas phase   12, 121
gels   93
George A. Olah   VI
George Bernard Shaw   193
gluconic acid   74
glucose   74
Goethe   v, 3, 15, 59, 75, 79, 81, 87, 105, 113, 153, 161, 197, 205
gold chloride   99
gold sols   99, 101, 102, 103, 104
gold solution   101, 105
graphite plates   33
greek mythology   201
ground state   52, 191
gunpowder   163, 164
gunpowder smoke   195

h
H_2 157
H_2O 25, 165
H_2O_2 49, 51, 52, 61, 84, 137, 138
$H_2Si_3O_7$ 95
H_2SO_4 27, 75, 91, 95, 125, 129
H_3PO_4 170
hair   2
Hans Magnus Enzensberger   85
$HAuCl_4 \cdot xH_2O$ 97, 99, 101, 103, 105
HCl   43, 63, 87, 133, 134, 171, 178
heart   2
heavy water   25
Heinrich Caro   61
Heinrich Roessler   35
Henry Cavendish   29
Heumann   29, 115

Hermann Hesse   133
hexa-aqua cations   89
hexane   165, 166
$HgCl_2$ 53, 123
HgO   183
HNO_2 75
HOCl   87
Hofmann electrolysis apparatus   55
human body   2
hydration enthalpy   7, 8
hydrazine hydrate   105
hydrocarbons   19
hydrochloric acid   35, 43, 44, 63, 85, 86, 95, 110, 133, 171, 177
hydrogen   31, 33, 34, 35, 37, 38, 55, 56, 79, 157
hydrogen bond   25
hydrogen disulfite   65
hydroxide   123
hypochlorite   28, 77, 78, 170

i
Icarus   201
ice   5
ice block   15
ice water   5
ideal gas   12
Immanuel Kant   39
indicator   92
indigo carmine   207
iodate   54
iodine   54, 83, 84, 115, 116
iodine-starch reaction   116
iron ball   5, 6
iron tube   197
iron(III) chloride   90
iron(III)thiocyanate   89
iso-propanol   139, 140

j
Johann Nestroy   69
Johannes Kepler   83, 183
John Tyndall   101
Jöns Jakob Berzelius   53
Joseph Rieckert   181

k
$K_2[HgI_4]$ 124
K_2CO_3 99, 103
$K_2Cr_2O_7$ 129, 130, 131
Karl Valentin   191
KBr   189

KCl 198
KClO$_3$ 27
KI 115, 123, 124
kinetics 47
KIO$_3$ 53
Kipp's apparatus 35
KMnO$_4$ 83, 87, 125, 127
KNO$_2$ 151, 164
KNO$_3$ 151, 159, 201
KSCN 61, 62, 137

l

Laotse 97
lattice energy 7
Lavoisier 41
lead(II) chloride 41
Leonardo da Vinci 43
Leopold Ruzieka 41
Lichtenberg 5, 57, 89, 107, 119, 121, 141, 145, 173, 185, 203
Liebig 33, 67, 110, 133, 195, 217
Linus Pauling 25
liquid ammonia 80
litmus solution 65
liver 2
Louis Pasteur 93
Ludwig Börne 11
luminescence 62
luminol 51, 52, 61, 62, 137, 138
lycopodium powder 186

m

magnesia boat 157
magnesia stick 67, 68
magnesium 161, 175
magnesium powder 161, 173
magnesium ribbon 161, 173
Manfred Rommel 77
manganese(II) 83, 84, 125, 126
Marie Curie 37
Mariotte 91
Mark Twain 23, 173, 217
Mars 1
Marshall McLuhan 1
Maxwell–Boltzmann 199
medicine 217
melting point 15
mercury(II) dichloride 124
metallic silver 110
methyl alcohol 75, 167, 168
methyl red 11
methylene blue 73, 74, 207, 209, 210

Mg 174
Mg$_3$N$_2$ 161, 162
MgO 161, 173, 174
Michael Faraday 101, 139
microscope 93
milk of lime 28
Millon's base 124
MnO$_4^-$ 83
molar mass 12
molecular mass 199
molecular velocity 37
molybdenum blue 63
Münchhausen v

n

n-hexane 83
n-propanol 139
N$_2$ 195
N$_2$H$_4$ 105
N$_2$O$_3$ 75
Na$_2$CO$_3$ 51, 89
Na$_2$O$_2$ 3, 4
Na$_2$S$_2$O$_4$ 47
Na$_2$Si$_3$O$_7$ 95
NaCl 159
NaH$_2$PO$_4$ · 2 H$_2$O 57
NaHSO$_3$ 53
NaNH$_4$HPO$_4$ 67
NaNO$_2$ 75
NaOCl 78
NaOH 61, 62, 69, 73, 91, 109, 137
Nessler's reagent 123, 124
NH$_3$ 55, 56, 80, 162, 178
(NH$_4$)$_6$Mo$_7$O$_{24}$ · 4H$_2$O 63
NH$_4$Cl 39, 55, 56, 178, 196
NH$_4$NO$_3$ 39
NH$_4$OH 55, 62
NH$_4$SCN 62, 89, 90
Nicolas Flamel 163
nitrogen 55
NO 75
NO$_2$ 75
Nobel laureate 61
Nobel Prize 93
Novalis 51
nylon rope 146

o

orthophosphoric acid 170
Oscar Wilde 146
oscillating chemical reactions 138
oscillating 62
osmosis 11, 12

osmotic pressure 11, 12
oxidation 151, 202
oxidizer 3
oxihydrogen 19, 157
oxygen 3, 31, 33, 58, 65, 66, 78, 83, 88, 146, 147, 151, 158, 170, 171, 183, 186
oxygen content 169

p

P_2 170
P_4O_{10} 170
Paracelsus 23, 29, 71
particle diameter 93
particle size 111, 181
$PbCl_2$ 41, 42
pecan kernel 149, 151
pentachloroethane 197, 198
pentane 141, 146, 147
pentane–air mixture 142
peppermint tea 49, 88
permeable 11
Pfeffer's cell 11, 13
pH value 105, 139
phenolphthalein 20, 139, 166, 19
phosphorus 27, 28, 67
phosphorus oxides 170
phosphorus oxoacids 170
photosensitizer 49
physiology 217
plasma 57
plasma state 57
plasmolysis 12
platinum 29, 31
platinum sponge 29
polycondensation 166
polyethylene terephthalate 147
polysilicic acids 95
ponceau 4R E 124 47, 77, 213, 215
potassium 19, 20, 123, 189, 197, 198, 203
potassium carbonate 99
potassium chloride 198
potassium nitrate 164, 202
potassium permanganate 125
pyrite 163
pyrolysis 181
pyrotechnic 145

r

radiation 202
Ralph Lippmann 19

Ralph Waldo Emerson 41, 97
Raymond Chandler 207
re-gelation 15, 16
red gold 97
redox reactions 73
rhodamine B 51, 52
Richard Wagner 61
Richard Willstätter 19
Richard Zsigmondy 93
rust 171

s

S_2Cl_2 196
S_4N_4 195, 196
salt melt 151
salt pearl 67
Saturn's rings 181
Schiller 63
SCl_2 196
sebacic acid dichloride 165, 166
self-organization 113
semi-permeable 12
semi-permeable membrane 11
silica gel 95
silver electrodes 107
silver nitrate 110, 111
silver residues 134, 214
silver sol 107, 109, 110, 111
smoke rings 177, 179
SO_2 163
sodium 3, 19, 20, 21, 79, 80, 139, 140, 203
sodium chloride 2, 5, 11, 43, 166
sodium dihydrogen phosphate 57
sodium fluorescein 51, 52
sodium hydroxide 30, 31, 35, 91, 110, 119, 126, 166, 171
sodium hypochlorite 47
sodium peroxide 4
sodium silicate 95
solid state 5
sols 93
solution enthalpy 7
solvated electrons 79, 80
sparks 142, 153
starch 53, 115, 116
status nascendi 35
steel wool 171
styrofoam ball 141, 142
sugar 17
sugar coal 17, 18
sulfur 65, 66, 164, 195

sulfur dioxide 65, 91
sulfuric acid 17, 27, 29, 30, 33, 34, 91, 125, 130
sunlight 202
surface 93

t
tannin 97, 111
tartaric acid 69, 70
teeth 2
tertiary amines 19
tetrakis(dimethylamino)ethane 191
Thales 1
thiocyanate 89
Thomas Alva Edison 137
Thomas Graham 37
tomato juice 87
toxic 129
transformer 34
transition 5
1,2,5-trihydroxybenzene 85, 86
Turquet de Mayenne 29
Tyndall effect 101

u
unbleached paper 193
uv light 193

v
van't Hoff 11, 27
vitriol 29
voltmeter 159

w
wash bottle 3, 38
Wassily Kandinsky 205, 209
water 1, 2, 3, 15, 17, 19, 20, 23, 25, 27, 38, 53, 57, 9, 105, 107, 125, 137, 169, 171, 181, 193, 203, 209
water blue 71
water content 1
water percentage 2
Werner Heisenberg 79
white phosphorus 169
Wiener xiii
Wilhelm Busch 165
Wilhelm Ostwald 141, 199
Willa Cather 43
WO_3 35
Wöhler 109, 153
Wolfgang Oswald 93
wood shavings 3
wooden splint 30, 56

z
zinc 30, 35, 110

Related Titles

Roesky, H. W., Möckel, K., Russey, W. E., Mitchell, T. N.

Chemical Curiosities – Spectacular Experiments and Inspired Quotes

354 pages with 65 figures
1996
Hardcover
ISBN: 978-3-527-29414-5

Bell, H. B., Feuerstein, T., Güntner, C. E., Hölsken, S., Lohmann, J. K.

What's Cooking in Chemistry?

243 pages with 149 figures
2003
Hardcover
ISBN: 978-3-527-30723-4

Greenberg, A.

From Alchemy to Chemistry in Picture and Story

664 pages with ((keine Angaben in PIV)) figures
2007
Hardcover
ISBN: 978-0-471-75154-0

Djerassi, C., Hoffmann, R.

Oxygen

128 pages with 4 figures
2001
Softcover
ISBN: 978-3-527-30413-4

Edwards, S. A.

The Nanotech Pioneers

257 pages with 46 figures
2006
Hardcover
ISBN: 978-3-527-31290-0